KU-247-491

Global Sustainability
The Impact of Local Cultures

A New Perspective for Science and Engineering,
Economics and Politics

Edited by
Peter A. Wilderer, Edward D. Schroeder, Horst Kopp

WILEY-
VCH

WILEY-VCH Verlag GmbH & Co. KGaA

Edited by

Prof. i. R. Dr.-Ing. Dr. h. c. Peter A. Wilderer
Institute of Advanced Studies on Sustainability
European Academy of Sciences and Arts
c/o Technical University Munich
Lehrstuhl für Wassergütewirtschaft
Arcisstraße 21
80333 München
Germany
peter@wilderer.de

Prof. Dr. Edward D. Schroeder
University of California
Department of Civil & Environmental Engineering
1 Shields Avenue
Davis, CA 95616
USA

Prof. Dr. Horst Kopp
Friedrich-Alexander-Universität
Institut für Geographie
Kochstr. 4/4
91054 Erlangen
Germany

All books published by Wiley-VCH are carefully produced. Nevertheless, authors, editors, and publisher do not warrant the information contained in these books, including this book, to be free of errors. Readers are advised to keep in mind that statements, data, illustrations, procedural details or other items may inadvertently be inaccurate.

Library of Congress Card No.: Applied for
British Library Cataloging-in-Publication Data: A catalogue record for this book is available from the British Library.

Bibliographic information published by
Die Deutsche Bibliothek
Die Deutsche Bibliothek lists this publication in the Deutsche Nationalbibliografie;
detailed bibliographic data is available in the internet at http://dnb.ddb.de.

© 2005 Wiley-VCH Verlag GmbH & Co. KGaA, Weinheim

All rights reserved (including those of translation in other languages). No part of this book may be reproduced in any form – by photoprinting, microfilm, or any other means – nor transmitted or translated into a machine language without written permission from the publishers. Registered names, trademarks, etc. used in this book, even when not specifically marked as such, are not to be considered unprotected by law.

Printed in the Federal Republic of Germany.
Printed on acid-free paper.

Cover Design   SCHULZ Grafik-Design, Fußgönheim
Typesetting   Manuela Treindl, Laaber
Printing   betz-druck gmbh, Darmstadt
Bookbinding   Litges & Dopf Buchbinderei GmbH, Heppenheim

ISBN 3-527-31236-6

Table of Contents

Global Sustainability. Edited by P. A. Wilderer, E. D. Schroeder, H. Kopp
Copyright © 2005 WILEY-VCH Verlag GmbH & Co. KGaA, Weinheim
ISBN: 3-527-31236-6

Preface

In the year 2000, the European Academy of Sciences and Arts (EASA) in collaboration with the Bavarian Research Centers BayFORREST and FORAREA started a series of workshops to explore, by means of a cross-disciplinary and inter-cultural discourse, the interrelationships between technological advances, cultural heritage and religious concerns. The aim of these workshops was to assist in the enforcement of the general Sustainability Development (SD) guidelines, and in the realization of the Millennium Development Goals (MDG) of the United Nations. The basic question underlying the discussions was: What can scientists, and what can engineers do to turn words into action?

Development of technology readily acceptable by people living under the cultural, ecological and economical conditions specific to the very geographical area is assumed to be highly important. To be able to implement sustainable development measures people need to have ready access to fresh air, clean water and safe food in sufficiently high quantity. Pollution of atmosphere, soil and natural water bodies must be prevented or mitigated to provide and maintain sound living conditions for present and future generations. Education must be made accessible for all people irrespectively of gender, social status and wealth, and irrespectively of the local conditions under which people use to live, conditions prevailing in a metropolitan or a rural area in Europe, in Asia, Latin America, Africa or wherever. It appears that one cannot be achieved without achieving the other: Winning the Millennium Development Goals is a prerequisite of making progress in sustainable development.

The workshops were organized to address those needs, and to make clear that cross-cultural and cross-disciplinary approaches are required to satisfy basic requirements of man and nature. The first of these workshops was held on October 28 and 29, 2000 at the Wildbad-Kreuth resort in the heart of the Bavarian Alps. The second workshop took place at the monastery "Kloster Banz", located in the North of Bavaria, Germany, on February 24 and 25, 2003.

36 scientists, engineers and entrepreneurs from 4 continents and 12 countries attended the workshop at Kloster Banz. Prior to the meeting overview articles were produced by 14 of the participants, and distributed among the attendees. With this information background an intensive dialogue began immediately. Except of two short introductory speeches, formal oral presentations were not given. At the beginning of the discussion session the various aspects of sustainability and sustainable development, culture and cultural diversity, needs and desiderates were discussed from a more general perspective. It was decided to focus during the further discussion on four specific areas:

A: World cultures and world religions
B: Poverty and economical development
C: Global and temporal dimension of sustainability
D: Technology, conflict and sustainability

Global Sustainability. Edited by P. A. Wilderer, E. D. Schroeder, H. Kopp
Copyright © 2005 WILEY-VCH Verlag GmbH & Co. KGaA, Weinheim
ISBN: 3-527-31236-6

In the following session, the participants split into four groups and discussed these topics thoroughly. The conclusions drawn were then presented to the plenum during the following session and finally summarized. It was also decided to write, as a joint effort, a book focusing on the cultural dimension of sustainable development, based on the abstracts formerly produced by various participants. This book is the result of the workshop and many bilateral discussions held afterwards.

September 2004

Peter A. Wilderer
Edward D. Schroeder
Horst Kopp

List of Contributors

Josef Bugl is Dean of the Class Technical and Environmental Sciences of the European Academy of Sciences and Arts. Between 1960–1972 he has held several research positions with Euratom. From 1973–1992 he has served as director with BBC Mannheim, responsible for developments in nuclear technology. From 1980–1987 member of the German Bundestag, spokesman for R and D of the Christian Democrats' caucus and chairman of the enquiry commission for technology assessment. He has been the initiator of the Academy for Technology Assessment in Stuttgart and chairman of the advisory board. From 1990–1998 honorary professor for technology assessment and environment at the Technical University of Chemnitz. Member of several boards and advisory committees.

Contact:
Prof. Dr. Josef Bugl
Elisabeth von Thadden Str. 7
68163 Mannheim
Germany
josefbugl@aol.com

Martin Faulstich became professor of waste management at the Technische Universitaet Muenchen in Garching. Since 2000 he is member and since 2001 chairman of the executive board of ATZ Development Center in Sulzbach-Rosenberg which is engaged in processes and materials for energy systems. Since 2003 he is also director of the new Institute of Technology of Biogenic Resources at the Technische Universitaet Muenchen in Straubing. In addition he serves as chairman of BayFORREST Bavarian Research Network on Waste Management, editor of the journal Wasser & Abfall, and member of numerous professional committees.

Contact:
Prof. Dr. Martin Faulstich
Institute of Technology for Biogenic Resources
Technical University of Munich
Am Hochanger 13
85354 Freising
Germany
m.faulstich@bv.tum.de

Global Sustainability. Edited by P. A. Wilderer, E. D. Schroeder, H. Kopp
Copyright © 2005 WILEY-VCH Verlag GmbH & Co. KGaA, Weinheim
ISBN: 3-527-31236-6

Martin G. Grambow *is civil engineer and executive officer in the Bavarian administration. He is head of the section for Water Management in Rural Areas and Torrents in the Bavarian Ministry of Environment, Health and Consumer Protection. In 1999 he founded in Hof the Bavarian office for Technology Transfer Water and supports as senior advisor projects in the water sector in East Europe, Asia, South America and North Africa. He is Vice President of the International Research Society Interpraevent and member of the Bavarian Academy "Ländlicher Raum" (Rural Areas) in Munich.*

Contact:
Dipl.-Ing. Martin G. Grambow
Bavarian State Ministry of Environment
Health and Consumer Protection
Rosenkavalierplatz 2
81925 Munich
Germany
martin.grambow@stmugv.bayern.de

Armin Grunwald *is director of the institute for Technology Assessment and systems analysis (ITAS) at the research centre Karlsruhe and professor at Freiburg University since 1999. Since 2002, he is also director of the Office of Technology Assessment at the German Bundestag (TAB). Since 2003, in addition, he is speaker of the Helmholtz programme "Sustainable Development and Technology". By education, Armin Grunwald is physicist and philosopher. His current working areas are theory and methodology of technology assessment, philosophy of technology, philosophy of science, and approaches to sustainable development.*

Contact:
Prof. Dr. Armin Grunwald
Institute for Technology Assessment and Systems Analysis (ITAS)
Research Centre Karlsruhe
Hermann-von-Helmholtz-Platz 1
76344 Eggenstein-Leopoldshafen
Germany
grunwald@itas.fzk.de

Michael von Hauff *is Professer for International and Environment Economics at the University of Kaiserslautern/Germany since 1991. Before he was associated with the University of Stuttgart. He was a Visiting Professor at the University of Delhi. He is a member of many national and international research associations like the European Academy of Sciences and Arts. His main fields of research concentrate on problems of development and environment economics.*

Contact:
Prof. Dr. Michael von Hauff
University of Kaiserslautern
Gottlieb Daimler Str. 42/404
67663 Kaiserslautern
Germany
hauff@wiwi.uni-kl.de

Gloria Hauser-Kastenberg *is a litigation attorney and mediator; Instructor at UC Berkeley. Ms. Hauser-Kastenberg received her B.A. from Temple University and her J.D, from Southwestern School of Law. She is a litigation attorney and mediator handling a wide range of cases primarily in the following areas: toxic tort and environmental issues, workplace and discrimination issues and professional malpractice cases. Since 1998, her practice has focused on mediation in private practice and in affiliation with several Bar Associations in the San Francisco Bay Area and the San Francisco Superior Court Civil Mediation Panel. She has worked as a facilitator, trainer and consultant with the Center for Changing Systems, a Bay Area non-profit organization and consulting company which teaches and utilizes a non-linear and systems approach for decision-making and communication. She has taught dispute resolution skills at UCLA and is currently at UC Berkeley jointly teaching "Ethics and the Impact of Technology on Society" with her husband Professor Kastenberg.*

Contact:
Gloria Hauser-Kastenberg
1678 Shattuck Avenue 311
Berkeley, CA 94708
USA
ghauser@igc.org

Hans Georg Huber *heads a technology corporation which is located in Germany and which, based on many years of tradition, has focused on the treatment of sewage and drinking water. The development of modern and forward-looking technologies corresponding to the requirements of high performance and cost effectiveness on the one hand and being tailored to the demands of international target markets on the other hand, has always been considered as a primary task. This also includes that financial feasibility, easiness of operation, ecological compatibility and economic practicability are taken into consideration. In the light of these aspects, solutions have been elaborated, developed and made available throughout the world.*

Contact:
Dipl.-Ing. Hans Georg Huber
Hans Huber AG
Industriepark Erasbach A1
92334 Berching
Germany
hgh@huber.de

William E. Kastenberg *is currently the Daniel M. Tellep Distinguished Professor of Engineering at the University of California, Berkeley. Dr Kastenberg is a member of the U.S. National Academy of Engineering, and is a Fellow of the American Association for the Advancement of Science and the American Nuclear Society. He has won distinguished teaching awards from the American Society for Engineering Education, and the Engineering Graduate Students' Association at UCLA. His teaching and research interests are in the areas of risk assessment and risk management for complex technologies, nuclear reactor safety, environmental risk analysis, and more recently, the study of ethics and the impact of technology on society.*

Contact:
Prof. Dr. William E. Kastenberg
4103 Etcheverry Hall
University of California at Berkeley
Berkeley, CA 94720
USA
kastenbe@nuc.berkeley.edu

His Royal Highness Prince El Hassan bin Talal,
President of the Club of Rome, Moderator of the World Conference for Religions and Peace, Chairman of the Arab Thought Forum

Horst Kopp *is head of the Department of Geography in the University Erlangen-Nuremberg where he holds the chair for Human Geography and Middle East Research. Besides he was chairman of the Bavarian Research Network FORAREA between 1995 and 2002 (research on area, business and culture). He serves as the editor of Jemen-Studien and other journals and he belongs the editorial board of Arab World Geographer.*

Contact:
Prof. Dr. Horst Kopp
Institute for Geography
Kochstr. 4/4
91054 Erlangen
Germany
hkopp@geographie.uni-erlangen.de

David Devraj Kumar *is Professor of Science Education at Florida Atlantic University. His research involves Science-Technology-Society studies, evaluation and policy. He has directed several funded projects including a national level status study of STS in all fifty states in the United States. He is co-editor of the book Science, Technology, and Society: A Sourcebook on Research and Practice (New York: Kluwer Academic/Plenum Publishers). He serves on the editorial boards of the Journal of Science Education and Technology, and the Review of Policy Research. Professor Kumar is a Fellow of the American Institute of Chemists.*

Contact:
Prof. Dr. David Devraj Kumar
Florida Atlantic University
College of Education
2912 College Avenue
Davie, FL 33314
USA
david@fau.edu

Amitabh Kundu *is currently Professor of Economics at Jawaharlal Nehru University, New Delhi. He has done Post Doctoral work as a Senior Fulbright Fellow at the University of Pennsylvania. He has been a Visiting Professor at University of Amsterdam, Maison des Sciences de L'homme, Paris, University of Kaiserslautern and South Asian Institute Heidelberg, Germany. He has done International Consultancies for UNDP, UNESCO, UNCHS, ILO, Government of Netherlands, University of Toronto, Sasakawa Foundation etc. He has worked as Director at various institutes such as National Institute of Urban Affairs, Indian Council of Social Science research and Gujarat Institute of Development Research. Currently he is in the Editorial Board of Manpower Journal, Urban India, Journal of Educational Planning and Administration, Indian Journal of Labour Economics. He has about twenty books and two hundred research articles, published in India and abroad, to his credit. His recent books (edited) are "Inequality, Mobility and Urbanisation: China and India", "Informal Sector in India" and "Poverty and Vulnerability in a Globalising Metropolis: Ahmedabad".*

Contact:
Prof. Dr. Amitabh Kundu
Jawaharlal Nehru University
Center of Studies of Regional Development
New Delhi 67
India
amit0304@mail.jnv.ac.in

Dimitris Kyriakou *holds a Bachelor of Science degree in electrical engineering & computer science, as well as Master's and Ph. D. degrees (dissertation on international economics issues) all from Princeton University. He works at the European Commission's Institute for Prospective Technological Studies, where his main tasks include those of the editor of the Institute's technoeconomic review, The IPTS REPORT, as well as those of the Institute's chief economist.*

Contact:
Dr. Dimitris Kyriakou
Institute for Prospective Technological Studies
Joint Research Centre
European Commission c/Inca Garcilaso s/n
Edf. EXPO
41092 Seville
Spain
dimitris.kyriakou@cec.eu.int

Hartmann Liebetruth *is professor of Managerial Economics at the Department of Print and Media Technologies at the University of Wuppertal. Since 1994 he is also the chairman of the »International Circle of Educational Institutes for Graphic Arts – Technology and Management«. He is holding regularly international seminars and workshops on developing new strategies for the print and media industry in its process of transition from a manufacturing to a service providing industry and was the co-author of the study »Future of Print & Publishing – Opportunities in the Media Economy of the 21ˢᵗ Century«.*

Contact:
Prof. Dr. Hartmann Liebetruth
Bergische University of Wuppertal
Department of Print & Media Technology
Rainer Gruenter Str. 21
42119 Wuppertal
Germany
liebetruth@kommtech.uni-wuppertal.de

Reinhard M. Mosandl *is the Head of the Chair of Silviculture and Forest Management at the Technische Universität München. In Germany his research activities are focused on natural regeneration in mixed mountain forests and on conversion of pure stands into mixed stands. On the international level he has supervised many PhD-theses aimed at sustainable management of natural forests. As chairman of the Section Silviculture within the German Association of Forest Research Organisations he has initiated several common research projects and publications. In addition to his scientific activities he is responsible for the management of the university forest where he combines scientific knowledge with practical experience.*

Contact:
Prof. Dr. Reinhard Mosandl
Chair of Silviculture and Forest Management
Technical University of Munich
Am Hochanger 13
85354 Freising
Germany
mosandl@wbfe.forst.tu-muenchen.de

David Norris *has for more than 25 years worked as an international educator, consultant, facilitator and coach in the United States, Canada, Australia, Europe, India and the Middle East. In 1967 he was awarded a Fulbright Fellowship to study Comparative Literature at the Free University of Berlin, He received his doctorate from Columbia University in New York City, later becoming a member of that faculty and then an Assistant Professor of Literature at City College of New York. Shifting from the academic to the business world, since 1979 he has been working with individual managers and project teams, in both English and German, to design and implement organizational cultures that provide a basis for non-linear thinking resulting in fundamental systems change. Currently he is developing and implementing a new understanding of risk perception in corporate culture.*

Contact:
Dr. David Norris
Allmendsberg 27
79348 Freiamt
Germany
dnorris@t-online.de

Ortwin Renn *is Professor of Environmental Sociology at the University of Stuttgart and Director of the non-profit company DIALOGIK, a research institute for the investigation of communication and participation processes in environmental policy making. He also chairs the European Center for Interdisciplinary Risk Research, Governance and Sustainable Technology, a research unit within the University of Stuttgart. From 1993 to 2003, he directed the Center of Technology Assessment in Stuttgart, Germany, a public foundation devoted to the study of the societal impacts of technological and social change. He has conducted a large number of research projects on risk perception, risk management, environmental policy analysis, conflict resolution, technology assessment, and energy planning. His current research includes environmental economics and sociology, risk perception, risk analysis and risk communication, regional concepts of sustainable development, citizen participation in risk management, attitudes towards technology, and social movements.*

Contact:
Prof. Dr. Ortwin Renn
University of Stuttgart
Department of Sociology II
Seidenstr. 36
70174 Stuttgart
Germany
ortwin.renn@soz.uni-stuttgart.de

Monica I. Renner *works as Professor of Administration at the Economic Science School of the University of Buenos Aires, Argentina, since 1988. She is in charge of the Center of Scientific and Technical Studies (C.E.C.y T.), acts as evaluator since 1994 for the Argentine National Quality Award and has authored various books and over 100 articles published. On invitation of the German Academic Exchange Service she visited the environmental management groups at the Cologne University and at the TU Munich. In 2003 she spent some time at the University of Queensland, Australia, as a visiting professor.*

Contact:
Prof. Dr. Monica Renner
University of Buenos Aires
Ada Elflein 3761 11A (B1637 AMG)
Buenos Aires
Argentina
monicarenner@yahoo.com

Dietmar Rothermund *is Professor emeritus of South Asian History, South Asia Institute of Heidelberg University. He is a Fellow of the Royal Historical Society, London, and Chairman, European Association of South Asian Studies. His major publications in English are Government, Landlord and Peasant in India (1978), Mahatma Gandhi. An Essay in Political Biography (1991), India in the Great Depression, 1929–1939 (1992), An Economic History of India (2nd rev.ed. 1993), The Global Impact of the Great Depression, 1929–1939 (1996), (with Hermann Kulke) A History of India (4th rev.ed. 2004).*

Contact:
Prof. Dr. Dietmar Rothermund
Oberer Burggarten 2
69221 Dossenheim
Germany
dietmar.rothermund@t-online.de

Werner Schenkel *held various leading positions at the German Environmental Agency since this institution was founded in 1974. First, he directed the division of water and wastewater management. Later he was in charge of the department of observation and assessment of the environment, and finally of the department of sanitation, ecology, water, soil and air. In 2003, he retired from his position but is still active as a senior advisor. He authored and co-authored more than 20 publications and serves as editor of a handbook on waste management.*

Contact:
Prof. Dr. Werner Schenkel
Franklinstr. 1
10587 Berlin
werner.schenkel@gmx.de

Edward D. Schroeder *is Professor Emeritus of Civil and Environmental Engineering at the University of California, Davis where he has been a member of the faculty since 1966. Professor Schroeder specializes in biological process engineering and has contributed to the understanding of kinetics and stoichiometry of both liquid and vapor phase treatment systems. He has authored or co-authored over 150 publications and three books on wastewater treatment, waste gas treatment, and water quality.*

Contact:
Prof. Dr. Edward Schroeder
Department of Civil & Environmental Engineering
University of California
1 Shields Avenue
Davis, CA 95616
USA
edschroeder@ucdavis.edu

Gabriele Weber-Blaschke *is researcher and lecturer at the Institute of Technology of Biogenic Resources at the Technische Universität München, Germany. After studying forest science and environmental engineering she made her PhD in forest science and worked as forest engineer in state and private forests. Back at the University she is performing projects involving sustainability, interdisciplinary and international cooperation and is teaching courses about waste and resource management. She published around 40 national and international papers in the fields of sustainable forest and resource management. Her main research interests are focused on forest and industrial ecology, management of material flows and indicator systems for sustainability.*

Contact:
Dr. Garbiele Weber-Blaschke
Institute of Technology for Biogenic Resources
Technical University of Munich
Alte Akademie 10
85354 Freising
Germany
gabriele.weber-blaschke@wzw.tum.de

Raoul Weiler *is engineer in Chemistry and Agro-industries and doctor in Applied Biological Sciences (1966) of the University of Leuven, Belgium. Before joining the chemical industry in 1970, he was a postdoctoral fellow at the University of North Carolina at Chapel Hill, the Catholic University of America, Washington DC and of the Centre National de la Recherche Scientifique, Paris, at the Université Paris V. In 1989 he was elected president of the Royal Flemish Engineers Association (K VIV) and founded a Working Party on Science, Technology and Society. At present he holds a teaching position at the University of Leuven, dealing with the relationship between technology and society. In 1997 he joined the Club of Rome of which he is now a member of the Executive Committee and is co-founder and President of the Brussels-EU Chapter.*

Contact:
Prof. Dr. Raoul Weiler
Prins Boudewijnlaan
113
2610 Antwerp
Belgium
raoul.weiler@skynet.be

Mirka Wilderer *works as a trainee of the Siemens AG working in the field of Power Transmission and Distribution. After studying Sociology in Regensburg, Germany, and Prague, Czech Republic, she graduated from the University of Bamberg, Germany in 2004, specializing in "European Integration and Global Society" and "International Management". Additionally, she completed the program of the Bavarian Elite Academy.*

Contact:
Mirka Wilderer
Moosweg 5
83727 Schliersee
Germany
mirka@wilderer.de

Peter A. Wilderer *is in charge of the Institute on Advanced Studies on Sustainability, funded by the European Academy of Sciences and Arts. He teaches at the German Institute of Science and Technology, Singapore, in Industrial Ecology and Industrial Chemistry. Besides he serves as honorary professor at the Advanced Wastewater Management Center of the University of Queensland, Australia. Between 1991 and 2004 he held the chair of Water Quality Control and Waste Management at the Techische Universitaet Muenchen. In 2003, he received the Stockholm Water Prize. He has made major contributions to the development of novel technologies applicable for advanced wastewater treatment. Professor Wilderer has authored and co-authored over 300 scientific publications. He serves as editor of the journal Water Research, and acts as editor-in-chief of the journal Water Science & Technology.*

Contact:
Prof. i. R. Dr.-Ing. Dr. h. c. Peter A. Wilderer
Institute of Advanced Studies on Sustainability
European Academy of Sciences and Arts
c/o Technical University Munich
Lehrstuhl für Wassergütewirtschaft
Arcisstraße 21
80333 München
Germany
peter@wilderer.de

Key Note Address

His Royal Highness Prince El Hassan bin Talal

President of the Club of Rome, Moderator of the World Conference for Religions and Peace, Chairman of the Arab Thought Forum

Today, rapid globalisation and technological 'progress' are dominating the daily lives of people in a seemingly uncontrollable manner. Whatever happened to the dictum that at the centre of development policy must be the 'human being'? Whatever happened to politics for the people – not 'hydropolitics' or 'petropolitics', but 'anthropolitics': politics where people matter? Shouldn't the end of development efforts be the betterment of human life; indeed, the betterment of *all* life? Surely, it is high time for a thorough critique of what policymakers, scholars, scientists and fieldworkers are achieving and *how* they are achieving it.

As the world continues to change at an ever greater rate, we urgently need to develop new mindsets – dare I say *axioms* – so as to integrate past experience and present activity under a set of universal principles. We can then be better prepared for the unexpected – the law of 'unintended consequences'. If we all realise that our behaviour (in developing as well as developed countries) influences others as well as ourselves, and if we accept the responsibility that therefore befalls us, we can all agree to a universal *code of conduct, an ethic of human solidarity*. There must be a call to establish mutually comprehensible terms of reference and value-systems in the interest of avoiding conflict and future loss. We are not in the business of assigning blame today; if we remember back far enough we realise that, as civilisations, we share responsibility for both 'good' and 'bad' global developments. A universal culture must now be agreed upon, established and abided by all members of our common globe, who face a common fate.

As Moderator of the World Conference on Religions for Peace, I call for a global code of conduct while promoting solemn respect for the various faiths and their interaction. Together with Dr. Hans Küng and Dr. Leonard Swidler, I stress that this code should have principles emphasising the association between theology and practicality; commonality; taking into account the Enlightenment tradition; embracing the principle of 'No Coercion'; upholding the right to proclaiming one's own religion; reconsidering the content of education; ensuring a free flow of information; looking courageously afresh at our own and each others' texts, heritage and history; developing a framework for disagreement; and producing a contextualised framework for further discussion and enquiry; as well as accepting responsibility for words and action at all levels. This recognises the political and economic dimensions of interfaith dialogue.

Global Sustainability. Edited by P. A. Wilderer, E. D. Schroeder, H. Kopp
Copyright © 2005 WILEY-VCH Verlag GmbH & Co. KGaA, Weinheim
ISBN: 3-527-31236-6

The degree to which our development efforts are successful or are resisted largely depends on the local context: perceptions, attitudes, religious or other beliefs, and behaviours of societies and individuals. These, in turn, depend on the community's history and surrounding circumstances: present and future. While none of these elements is stable, they must still be seriously taken into consideration not only to avoid a waste of resources but also to avoid drastic results. A better understanding of people's perception of their reality; their conceptualisation of God, nature, Man's role and 'fate'; and their feeling of belonging to the world at large – all these ingredients will surely facilitate designing suitable programmes. By 'suitable', I mean: useful as well as least disturbing to local, regional and global culture and natural environment.

As part of an anthropocentric approach, it is essential that we listen to what the human being wants. This sounds simple; yet it necessitates a healthy, functional civil society and an efficient, accountable and democratic infrastructure – to name a few prerequisites. And let us remember that before we begin to construct a civil society, we must first build a community.

Throughout history, societies have adapted to their local realities, finding the most affordable and appropriate means to arrive to least harmful consequences. Several examples exist of how local behaviour, when barred for different reasons, has actually resulted in the depletion of natural resources and direct harm to the people involved. The imposition of market forces on communities which previously were not connected with the global economy can have horrific consequences if not managed carefully and in concert with the community's own requirements and traditions. A terrible example of market impact was the destruction of the peoples of the Upper Amazon during the period 1894–1914 (the 'rubber boom'). The indigenous tribes suffered environmental destruction and also displacement, disease and murder as a consequence of bad treatment by the rubber companies, which led to the loss of 90% of their population. Several such examples, unfortunately, exist in our world today.

As emphasised in the Report of the Independent Commission on International Humanitarian Rights, which I had the honour to co-chair with Prince Sadruddin Agha Khan, at the core of indigenous cultures is their relationship with their land. They share a worldview which incorporates as its fundamental principle the custodial attitude to land and its natural resources. Indigenous peoples regard the land as a living entity entrusted to them for safekeeping and for passing on intact to future generations. I find the concept of 'multiple modernities' extremely valuable. Human values and global sustainability can be promoted by recognising that for many in the developing world, development is an attempt to 'indigenise' modernity rather than to 'modernise' traditional societies.

Concentrating on the economics of the situation alone is unhelpful. Earlier models for sustainable tapping of natural environments or 'extractive reserves' did not take into account the need of the local population to have control over what is happening to them and then to participate in a bottom-up process in improving their standards of living. It is not a question of 'money versus biodiversity', but one of engaging interactively with the local cultures to find out what can work according to their own exchange systems, territorial claims and methods for environmental maintenance. Small-scale development projects for economic subsistence have to be balanced with a good understanding of social and cultural needs. Communities want to have their own say about what affects their lives. This must be distinguished from the say of other cultures in the same country as well as that of the governments of these communities. A condition for a continuous and worldwide sustainable dialogue *within* and *among* cultures must be established.

As we speak of 'globally healthier' alternatives, let us keep in mind that poverty often dictates necessary evils due to a substantial lack of capabilities, means and awareness amongst the poor. We must talk about enabling the poor to cope with poverty as we expect their cooperation in what is good for the globe. Let us also keep the poor at the centre of our efforts – let us make them stakeholders in our development plans – as we realise that they are the first to pay the price to development mishaps and the advance of the powerful.

Different cultures share similar problems and, in many cases, similar backgrounds. Realities within borders are interlinked with conditions elsewhere. Behaviours *within* strongly affect realities outside. And since we all share a common globe with a common destiny, it is time we highlighted common values that call for the survival and sustainability of our lives. We need to realise a system whereby we can protect our traditions at the same time as encourage a unified global civilisation – in that we should and can exist as 'one world of ten thousand cultures'. It seems to me that we are very much in a position to realise that we are one civilisation, sharing basic human values which do not clash; and that we should grasp the opportunity for a quantum leap in the implementation of a global culture of cooperation.

This is why in the Club of Rome (which I have the honour to preside upon) we have added a fourth pillar to sustainability: that of culture, in addition to the social, ecological and economical dimensions. Human values and global sustainability mean that an alternative must be found to the imagined and feared hegemonic and homogenic globalisation process. In terms of the common standards of humanity and fundamental human rights, it is possible to perceive of a global ethic for all cultures, religions and faiths, together with multiple modernities.

Willingness to participate in constructive good will is a necessary foundational value. However, if countries decline to participate in a global process, or restrict the rights of their indigenous populations to have a say, the process will necessarily continue to develop in a nondemocratic manner. By the same token, if peoples and countries insist justly upon their right to move from a culture of existence to a culture of participation, they are working to democratise global processes in their own interests.

Multilateralism – and not bilateralism – is also a key in avoiding a bleak future. Organisations such as the UN, WIPO, and WTO can help encourage global cooperation and dialogue – in their policies, conferences and programmes. Interconnectivity has become the order of the day. I have encouraged worldwide e-conferencing and the launch of virtual universities. I have also participated in the initiation of several fora that facilitate knowing the 'other'. An example is the Parliament of Cultures where men and women from all over the world will participate in a conversation focused on two themes: education and the media. During the preparation process in Ankara a few months ago, I announced that through such a parliament I hope to promote local values, sustainable practices consistent with global reality, and an access to knowledge for all human beings.

Cultures and faiths may differ in some aspects and diversity must be respected. Within this rich diversity, however, lies a firm foundation of common values. Values that would never call for a destruction of what is good, or a disastrous end to what must be sustained: life. Let us build upon this common foundation – our 'global commons' – for the sustainability of our common globe.

1 History and Mandate of Sustainability: From Local Forestry to Global Policy

Gabriele Weber-Blaschke*, Reinhard Mosandl**, and Martin Faulstich***

* Institute of Technology for Biogenic Resources, Technical University of Munich, Alte Akademie 10, 85354 Freising, Germany, gabriele.weber-blaschke@wzw.tum.de
** Chair of Silviculture and Forest Management, Technical University of Munich, Am Hochanger 13, 85354 Freising, Germany, mosandl@wbfe.forst.tu-muenchen.de
*** Institute of Technology for Biogenic Resources, Technical University of Munich, Am Hochanger 13, 85354 Freising, Germany, m.faulstich@bv.tum.de

1.1 Sustainability: Key Word in the Today's Policy Discussion

Since the environmental world summit 1992 in Rio de Janeiro "Sustainability" has been the ideal not only for the environmental policy but also for all other political fields. The Agenda 21 concept of "Sustainable Development" is presented as a fixed mandate for all members of the United Nations.

Although the term "sustainability" was used in the last ten years in nearly every context and at almost every opportunity, there are serious doubts whether this key word is universally understood. For example, only a very small proportion of the German population truly grasps the meaning of the terminology [1]. Even at the conferences of "Wildbad Kreuth" (2000) and "Kloster Banz" (2003), dealing with the topic "sustainability" and setting the basis for this book, no consensus was reached concerning the meaning of this key word. Some participants attempted to explain the word by using the expression "to keep something on a certain level." For others sustainability meant "the reconciliation of society's development goals with its environmental limits over the long term." A further group held that the word "characterizes an economy in which humans can live on the interest of the natural capital without depleting the capital itself." Some participants wrongly believed that the term "Sustainability" is an artificial expression and has translated into German language with "Nachhaltigkeit", which is also thought of as an artificial word.

Global Sustainability. Edited by P. A. Wilderer, E. D. Schroeder, H. Kopp
Copyright © 2005 WILEY-VCH Verlag GmbH & Co. KGaA, Weinheim
ISBN: 3-527-31236-6

The situations in "Wildbad Kreuth" and "Kloster Banz" were not very different from many other international conferences and political discussions about sustainability. At the conclusion of the conferences it became apparent that there was neither a consensus about the content of this term nor about the strategies and actions which lead to sustainability. Neither within scientific nor political nor industrial circles does an agreement about the meaning of sustainability exist. In such a situation there is a great danger that an idea will become indeterminate and meaningless.

It is widely supposed that "Sustainability" was created first by the international environmentalists of the 1980s, but this is not the case. In actual fact, the principle of sustainable use of wood and forests was developed in German forestry approximately 300 years ago. Forestry in the 19th Century was already seen as a model for other economic sectors because of foresters' actions aimed at realizing sustainability [2]. Therefore, it is the objective of this article to go back to the roots of this supposedly modern word, "Sustainability," and attempt to extract its meaning from its history. The transfer of sustainability into other economic sectors must begin with an analysis of the original concept. After the description of the historical context and the mandate of sustainability, some examples will be presented in order to show how ideas and models towards sustainability might be transferred from forests and forest management into actual industrial and societal systems and actions.

1.2   History and Definitions of "Sustainability"

1.2.1   Development of the Technical Terminus in Forestry

The term "Forstliche Nachhaltigkeit" („Forest Sustainability") was presumably used for the first time in German forestry, though the ideas of sustainable forest management were also developed in other regions of Middle Europe, especially in French Forestry during the 16th and 17th century [2, 3].

Until now, the oldest documented application is found in the 1713 publication of a comprehensive German technical book about forestry by von Carlowitz [2, 4]. In his book "Sylvicultura oeconomica" von Carlowitz [5] coined the term "nachhaltende Nutzung" („sustainable use") because the traditional expression, "pflegliche Nutzung" („careful use"), appeared inadequate for his intention to manage forests in a wise manner in the long term [3]. While formulating his concept, von Carlowitz was inspired by strong religious beliefs: for him, nature was not a pure deposit of resources, but rather an act of the omnipotence of God. Therefore it was recommended to treat the gifts of God – mines and forests, for instance – gently. When faced with the possibility of the source of richness in Saxony, the silver mines in Freiberg, running dry – not because of the exhaustion of the mines, but because of the lack of wood for the melting process – it was absolutely necessary to find a long-term supply of wood. At the turn of 18th century a wood shortage was expected after forests were destroyed by clear cutting, excessive felling, removal of stumps and other overuses in the Middle Ages in order to support ship construction and mining. In some newer publications the existence of a general lack of wood in Middle Europe at that time is questioned [6], but there is no doubt that in many local areas, like the "Erzgebirge" of Saxony, the forests were destroyed because of the overuse of wood for ore smelting. Von Carlowitz, who was responsible for the mines and the forests of this area, recognized the negative impacts of the ruinous exploitation of the forests and developed the principle "to use the wood carefully, so that equilibrium results between increment and use of the wood" [3].

The application of these ideas was favored in this time by the age of Enlightenment. The faith in the mankind's progress by the means of rational actions led to a strengthening of theoretical science and to an increase in forest research. Theoretical models of forest management for practical actions were developed and included in the formulation of "Forest Sustainability".

First, forest sustainability was related to economic aspects, conservation of forest area, wood production and sustained wood yield, and later also to maximization of the working capital or total financial forest yield. In the further development of forest management the concept of sustainability was transferred to the entire forest ecosystem with the objective to consider the ecosystem functioning as well as the use, protection, recreation and cultural functions for mankind.

For example, in 1997 Pro Silva [7], a forest association, declared, *"Sustainability in forest management must not only be directed at timber and other marketable commodities, but at the full range of functions of forest ecosystems. Such a broad approach to sustainability includes: The maintenance of biodiversity as referred to in Agenda 21 of the Rio Conference: species diversity, genetic diversity, spatial and temporal diversity in structure; The maintenance of protection of hydrology, soil and climate; The maintenance of the natural fertility, health and productivity of the forest, and where applicable their restoration; The ability of forests to meet people 's physical and spiritual demands."*

In general, every forest management model includes questions concerning the ecological basis, demands of society and technological feasibility [8]. Therefore the term "Forest Sustainability" contains the ecological, the social and the institutional/economic dimensions, as well as the temporary and the spatial dimensions as its bases for all human actions [2].

1.2.2   Application in Environmental Politics

In the 1970s and 1980s, discussions concerning resource use, economic growth and their relationship to impacts on human beings and their environment arose in the face of signs of global environmental damages and catastrophes. The basis for these debates was the so-called Meadows report entitled "Limits of Growth," written in 1972 for the "Club of Rome" under the direction of the German "Volkswagen-Foundation," which was the first study related to "Sustainable Development" in this context. In the same year the first worldwide environmental conference took place and in 1983 the United Nations World Commission on Environment and Development was founded with the aim of making progress in environmental politics.

The Commission achieved its goal with the publication of a report in 1987, in which the situation of the world was analyzed. In the now famous Brundtland report, the term "Sustainable Development" was defined as *"development that meets the needs of the present without compromising the ability of future generations to meet their own needs"* [9]. According to the concept defined by the Commission, sustainable growth considers ecological possibilities as well as social justice for the accomplishment of the basic needs of all people. The main aspect relating to ecology is the finite nature of resources and the limited capacity of the biosphere and ecosystems to handle human impacts. With social issues, the justice between generations and within every generation should be considered.

At the world summit 1992 in Rio de Janeiro the term "Sustainable Development" was spread around the globe. In Agenda 21, United Nations member countries were obliged to adopt the concept of "Sustainability", which was described as a three-dimensional model (Figure 1.1).

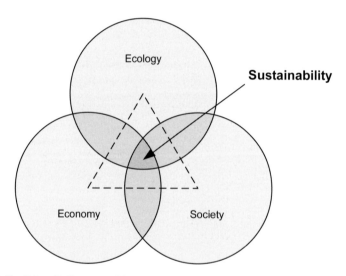

Figure 1.1 Traditional" Concept of Sustainable Development: Equivalence of the Three Dimensions Ecology, Economy and Society – An Anthropogenic View of Sustainability?

Sustainable development means not only the durable protection of the environment and resources, but also the achievement of social and economic goals. The categories of ecology, economy and social affairs are understood in an equivalent range [10]. In this model, sustainable development is also said to be taking place where all three sets of interests overlap in the center of the model.

However, the German Council of Environmental Advisors criticized this softened definition of sustainability [10], which is also used in the sustainability strategy put forth by the German government in 2002. From the Council's point of view, the three-dimensional model is related only to the Brundtland report's objective "accomplishment of the <u>basic needs</u> of everybody" [9]. Therefore, the Environmental Council understands the ideas of "Sustainable Development" as an ecologically-focused concept, in which social and economic cross sectional relationships play important roles because natural capital may be substituted by material or human and knowledge capital only in a low degree [10]. The "modern" concept of sustainable development is offered (Figure 1.2), where society and economy exist as a subset of the environment [11]. Human society is understood to be a part of the environment and therefore the environment must be conserved as the life basis for human beings. In addition, the economy is not a category existing for itself, but only a tool to regulate the activities of society. This philosophy might be characterized by its more bio-centric view (as seen by human beings, of course), whereas the three-category concept (Figure 1.1) centers on mankind and has takes a more anthropogenic view.

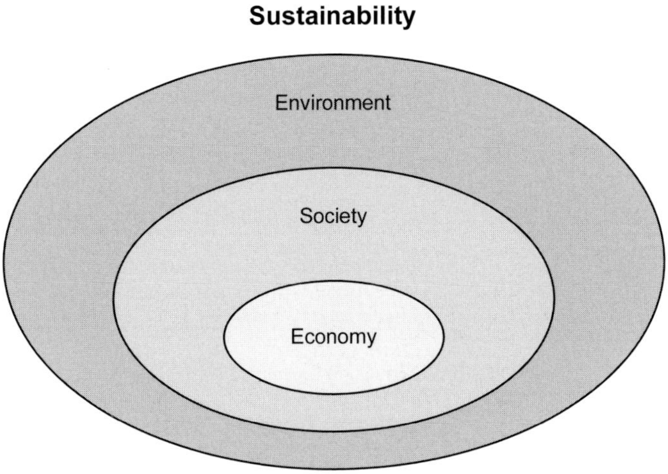

Sustainability

Figure 1.2 Modern" Concept of Sustainable Development Economy as only Instrument for the Society, Society as only Part of the Environment – A Bio-Centric View of Sustainability? (With changes according to Williams [11])

1.2.3   The Definition of "Sustainability" – A Social Negotiation Process on Local and Global Levels?

The history of "Sustainability" shows that there were two independent developments of the term: the development in forestry three hundred years ago and the younger development in environmental policy in the 1980's [2]. Nevertheless, the meaning of the contents seems similar, namely to consider ecological, economic and social issues.
But are we now clever enough to understand and to act towards sustainability not only alone, but also as society? That is in doubt because of the varying definitions that exist in different economic sectors and countries.
   The variety of definitions of "Forest Sustainability" in Germany alone depends on different conceptions and interests of the actors and the changing norm and value sys-

tems of the society. They also depend on different regions and times, as the developing processes of "Forest Sustainability" in Middle Europe and the United States of America show [2]. This demonstrates that the term is clearly affected by "time and place in response to prevailing social, economic and political conditions" (according to Lee 1990 cited in [2]).

The sophistication and content of indicator systems for sustainability, which should be used to analyze and control the development towards sustainability, are also related to the presence of influential political, geographical, economic and social conditions as comparisons of sustainability indicator systems in Germany, Hungary, Greece, and the USA have shown [12].

The social situation plays a great role in the interpretation of different points of view or different ways of behavior. Theoretically there exist many perceptions of reality and basic assumptions by actors in the relationship between human beings and nature [2].

Every religion or philosophy has a special ideology about the relationship between "God", human beings and nature [13]. Krieger describes the function of religion as to assert criteria of sense, truth, reality and basic convictions about the nature of the human beings and the order of the world as well as the highest values and norms of human behavior and to produce conditions for finding agreements.

Therefore we must conclude that there is not one exclusive definition of "Sustainability", but it is a social negotiation process that reflects the social circumstances and the power conditions in a specific region at a specific time [2]. Accordingly to Suda & Scholz [14] the result is always a compromise for the actors who do not inevitably work toward "Sustainable Development" considering all dimensions of sustainability.

1.3   Experiences from Forestry and Transfer into other Economic Fields – Possibilities and Limits

Actions and management toward forest sustainability are based on knowledge of forest ecosystems, their organisms and natural (and anthropogenic) environmental factors, which interact with material, energy and (genetic) information flows. Because of the long-term functioning and stability of such natural systems, their main material and energy flows model, especially concerning recycling, has been transferred into societal and industrial systems. A comparison between models of forest ecosystems (Figure 1.3) and of industrial ecosystems (Figure 1.4) and their management conditions will show the possibilities and limits of this transfer, particularly in actions towards sustainability and into other economic sectors.

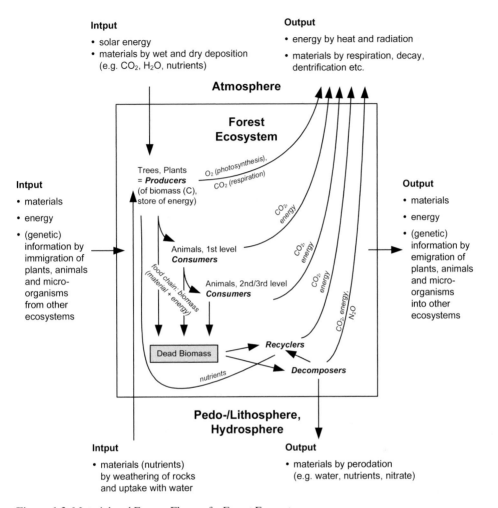

Intput
- solar energy
- materials by wet and dry deposition
  (e.g. CO_2, H_2O, nutrients)

Output
- energy by heat and radiation
- materials by respiration, decay,
  dentrification etc.

Atmosphere

Forest Ecosystem

Trees, Plants
= *Producers*
(of biomass (C),
store of energy)

O_2 (photosynthesis),
CO_2 (respiration)

CO_2 energy

Animals, 1st level
Consumers

CO_2 energy

food chain: biomass
(material + energy)

Animals, 2nd/3rd level
Consumers

CO_2 energy

CO_2 energy, N_2O

Recyclers

Dead Biomass

Decomposers

nutrients

Intput
- materials
- energy
- (genetic)
  information by
  immigration of
  plants, animals
  and micro-
  organisms
  from other
  ecosystems

Output
- materials
- energy
- (genetic)
  information by
  emigration of
  plants, animals
  and micro-
  organisms
  into other
  ecosystems

Pedo-/Lithosphere, Hydrosphere

Intput
- materials (nutrients)
  by weathering of rocks
  and uptake with water

Output
- materials by perodation
  (e.g. water, nutrients, nitrate)

Figure 1.3 Material and Energy Flows of a Forest Ecosystem

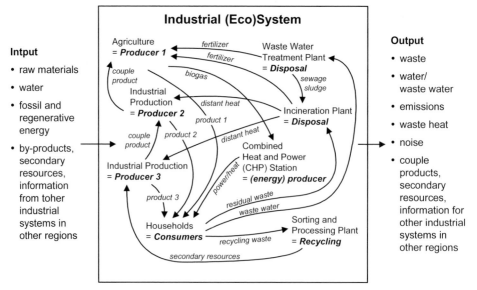

Figure 1.4   Material and Energy Flows in an Industrial (Eco)System
   a) The Product Based Approach including the Life-Cycle Philosophy
   b) The Geographical (Regional) Approach for a Recycling Network
       (With changes according to Korhonen [19])

1.3.1   Peculiarities of Forest Ecosystems and Forest Management

Forest ecosystems and their management can be classified according to the aforementioned dimensions of sustainability.

With regard to the *ecological dimension,* forest ecosystems are dynamic, open and long-term systems with (genetic) information, energy and material flows within the system and to and from other ecosystems and the environment, such as the atmosphere, pedo-/lithosphere, and hydrosphere (Figure 1.3, see [15]). The basis for the ecological, dynamic balance, a constant state between the different physical, chemical and biological interactions and energy, material and information flows in the ecosystems during a special time period, is the self-regulation of the ecosystems. In the long-term perspective forest ecosystems are stable systems that resist disturbances and which thoroughly include temporary or durable changes into the original organism pools or energy and material flows. Some examples include storm or fire events for short periods, climate changes during the ice-age, or for long periods in modern times. The sustainable management of forests must take into account that the production plant (forest) is nature and that the production of wood is executed by using the natural processes. The main advantage of the forest model is seen by many in the cyclic systems. On the one hand the producer – (consumer) – reducer – producer cycle within the ecosystem is respected, and on the other hand the C-cycle, also as related to the environment concerning, which leads to the CO_2-neutrality of forests and forestry [16].

Within the *economic and technical dimension,* several points should be emphasized. The first advantage is that wood is a renewable resource. Second, the forest ecosystem uses only "free" solar energy and develops functioning and ecologically-sound recycling processes and mechanisms for self-regeneration against damaging internal and external impacts. Therefore the forest ecosystem is permitted to have high energy use and loss within the food chain, as well as an intensive material conversion and metabolism. Additional differences between forestry and other economic sectors must also be considered. For management purposes it is necessary to maintain (growing) stock and to plan for uncertainties in market needs and prices in the long-term, such as maintaining different tree species in order to have a varied product supply. Additionally, the product is the educt at the same time („wood growths on wood"). Disturbances like storm or fire, which do not represent problems for the forest ecosystem in the long term (see above), may spell disaster for the short-term economic side of forest management.

Furthermore, the *social dimension* plays an important role. The forests must be able to handle different social pressures, such as wood/timber production, employment, protection and recreation. However, forests and their history influence the culture of the peoples who live and work in surrounding the area.

The *temporary and spatial dimensions* of forests and forestry are the very basis for the sustainable engagement of foresters. The long-term aspects of ecosystem processes, and especially wood production, lead to further consideration on local and regional levels, as well as on national and international levels.

These peculiarities of forests and forestry result in *long experiences for forest sustainability*, which may be transferred to other sectors, but not before proving the analogy and adoption of this concept. The development and application of sustainability in German forestry was promoted by forest education in several universities and forest schools since approximately 1750, when the students were lectured in technology as well as in natural, social, economic and political sciences. Additional practical applications and scientific research developed concepts within the principles of sustainability, such as the concepts of durable forests without clear-cutting (Möller 1920), multi-species, mixed forests (Gayer 1886), and natural forest management (e.g. ANW 1950) (see [17]).

1.3.2   Concept of Industrial Ecology

Due to the previously described advantages of natural ecosystems, a concept for an industrial environmental management called "Industrial Ecology," which transfers the principles of natural ecology, was developed. Consequently, industrial systems constructed as "Industrial Ecosystems" seem analogous to natural ecosystems (see [19]). The major problems in industrial and societal systems involve waste generation, pollutant emissions, and resource overuse that damages and destroys the environment, the life basis for human beings. Therefore the concept of "Industrial Ecosystem" may be understood as an attempt to use the natural recycling model in which the different actors cooperate by using each other's waste material and waste energy flow to minimize exploitation of virgin material and energy input as well as the waste and emission output from the system as whole.

This concept is based upon two approaches, which may be used separately or together.

One approach is the so-called product based approach (Figure 4a), in which the life cycle of a product, beginning from raw material extraction, is considered until the disposal of the product after its consumption or utilization. The information, material and energy flows along this supply and delivery chain, including suppliers, customers, stakeholders and co-operation partners are optimized [19]. Concepts like the international standard 'Life Cycle Assessment (LCA)' or the European approach 'Integrated Product Policy (IPP)' are tools for application and part in the management systems of the involved enterprises.

However, for the main parts of products, the production and end-consumption are geographically separated, which is a major driver of societal material flows. The result is high energy consumption. Additional tendencies toward problem displacement to other areas or to other enterprises may also be observed.

Therefore, a second approach, called the geographical approach (Figure 4b), is necessary to better manage the energy and material flows of a region [19]. In this concept a recycling network of all industrial actors in a special region is built up. The objective is also to use coupled products (= products that are generated as byproducts of generation of the main product) such as waste material and waste energy as resources with high value to reduce the virgin material consumption and energy input as well as the waste and emission output, and of course the related costs, too.

The goal of 'Industrial Ecology' is to address the problem of industrial society's 'throughput' material and energy flow and install contrasting material and energy flow models termed 'roundput', which means cyclical and cascading flows like those found in natural systems [20].

The basis for management of these systems, whether a forest ecosystem or an industrial (eco)system, is the knowledge about the material and energy flows, their amounts (analysis), their impacts (assessment) and their management instruments (development and execution of strategies), as described by Weber & Faulstich [21, 22]. However, this procedure includes not only analysis of the material and energy flows – the natural scientific and technique view, but also analysis of information flows and the interests and actor behavior – the economic and social scientific view.

1.3.3 Limits in Transferring the Forestry Model to other Economic Sectors

With the knowledge of forest ecosystems and their management, it seems logical to use their functioning, stability and sustainability to transfer their model to other economic sectors. However, the analogy of forest and industrial ecosystems has limitations, which are described and explained in the following examples.

Example 1: Use of finite resources
As described above, sustainability is created in forestry with the aim of using "the wood carefully, so that equilibrium results between increment and use of the wood". In the context of "non-regenerative," finite raw materials like mineral raw materials for the building sector, however, this term must inevitably include another strategy. With the basis of the three dimensions of ecology, economy, and social issues, the objective of a common environmental policy in this field is to ensure a local, economically and environmentally sound supply of mineral raw materials in the future.

For solutions of conflicts between raw material safeguarding, conservation of small and middle enterprises, and protection of space, resource, groundwater, soil and nature tools of environmental planning and of material flow management are necessary. The differentiated analysis of the raw and building material flows, including their energy consumption, is the necessary basis for the development of management strategies, and especially the substitution of raw materials by recycling of renewable materials plays an important role [23].

Example 2: Use of non-renewable, fossil energy stocks
Similar to mineral resources, non-renewable fossil energy stocks cannot be used in a sustainable way as within the forestry model. The alternative is the economical use so that the next generation will also participate in this supply, but whether it is also possible for the following generation is questioned. Furthermore, the use of the immobilized fossil energy sources by incineration leads to additional CO_2 emissions, to which climate change can probably be attributed.

The main differences in forest ecosystems are the use of solar energy and CO_2 neutrality. This is the main basis for sustainable forestry, including the wood industry [16].

Example 3: Limits of recycling
The forest ecosystem has a well functioning recycling system, although waste in the form of dead biomass is also generated and accumulates from time to time, as in the case of raw humus under a spruce or pine stand on an acid site.

In contrast, the transferred recycling system in the human economy is not yet perfectly developed. Collection, transport, and processing of the waste also require energy and resources. On the other hand, recycling circles run the risk of accumulation of pollutants and reductions in product quality (down-cycling) as the crises and discussions of BSE, sewage sludge and plastics recycling have demonstrated in the past. According to current experiences the recycling economy will contribute neither to resource nor energy conservation, but at the same time it will advance the distribution of pollutants and nutrients in the environment, increase the mobility by transports and have unknown social and ecological side-effects (Schenkel & Schramm 1998 in [22]). Therefore, waste avoidance must be given priority over waste recycling.

Example 4: Complexity of networking
Stable forest systems are based on well-functioning relationships and linkages between organisms and related material, energy and information flows, and their self-regulation after disturbances developed over a long period of evolution.

In industrial systems, organized relationships between the actors, such as delivery chains (Figure 4a) or recycling networks (Figure 4b), are preconditions to reduce environmental impacts and related costs. The numbers of different actors involved in an industrial (eco)system can be seen as diversity. It creates possibilities for increasing connectedness and cooperation in waste and by-product utilization within the system [20] but simultaneously increases the complexity of the network. This complexity results not only in unknown ecological effects but also in unpredictable relations between the actors because industrial systems are not biological systems but societal cultural systems [20]. The feedback of such actions could change material and energy flows as well as varying economic conditions like supply, demand for products, and related prices and the social situations of the actors (see chapter 2.3).

The reason for the stability and functioning of the forest ecosystem may be seen over the long-term of evolution time, which is approximately three hundred million years (since the Carbon period when the so called coal forests existed), whereas industrial systems cannot be optimally developed from the beginning of industrialization because of their young age of only around one-hundred years.

1.3.4   Approximation to the Ideal "Sustainability"

An important recognition is the fact that sustainability is an ideal that can never be fully achieved in reality. Even in forestry, the only economic branch, which is thought to be sustainable in some aspects, sustainability is limited. When human beings use a forest they alter its structure and composition. As a result, the forest is not in the same condition following the harvest as it was before the harvest, and the system is not sustainable in the sense of a constant state. This was already known in 1816 by the famous forester

Heinrich Cotta, who wrote in the introduction of his book "Anweisung zum Waldbau" („Instruction for Silviculture") [24], that the forests grow and exist in their best condition when no human beings and no forest sciences are influencing the forests. But people need wood for their livelihood and we are thus forced to harvest wood and change the status of the forests. The decisive point in this situation is the manner in which the condition of the forest is changed. Cotta wrote that a good forester disturbs perfect forests, whereas a bad forester ruins them everywhere.

Despite the fact that sustainability in a strong sense is an ideal and therefore beyond reach, the concept is helpful as a reference. Certain human activities can be ordered in their proximity to this ideal state. Without a doubt, good forestry operates very close to the ideal state of sustainability and it can therefore be taken as a standard for other economic sectors. On the other hand, it must be admitted that in branches working with non-renewable resources, sustainable development has some underlying limitations [25]. Thus, each branch must be considered and judged separately by specially developed sustainability criteria with their respective indicators. A further step is comparison among the branches. But the evaluation and assessment of the different focal points and problems, like different resource uses or environmental impacts, are very difficult. Nevertheless, in spite of major differences among various branches, there exist a few guidelines leading to an approximation of the ideal "sustainability":

The objective should be an *intelligent use of resources*. This can be achieved by punishment of lavish use or rewards for conservative use.

The production process should have *minor impacts on the environment*. Payment for the pollution of water, soil and air with toxic or undesirable substances could promote sustainability in the same way as rewards for clean production processes.

Renewable resources should be given preference over non-renewable resources. Products from renewable resources should become more attractive than products from non-renewable resources.

Taking these guidelines seriously has the following economic consequences: Energy and products from fossil resources will become more expensive. Transportation costs will increase significantly. Water, soil and air will no longer be free. But this increase of certain prices will result in a development towards sustainability.

1.4   Conclusion

Perfect sustainability is an ideal without a chance of realization. But there is a chance to shift production processes toward sustainability. Forests and forest management could be considered in this case as references, which are very close to sustainability. The preconditions for such a change must be created by a new economic framework.

The acceptance of the sustainability concept and also of the associated necessary economic changes requires a great deal of responsibility. There are strong doubts that this responsibility emerges from common sense. It is much more likely that the ethic or religious attitudes of people are the prerequisites necessary for the development towards a more sustainable world.

Acknowledgement

We are grateful for revisions suggested by Geoffrey Painter, University of California-Berkeley, USA.

References

[1]   M. Suda, F. Zormeier, Forst und Holz 2002, 10, 322–324.

[2]   H. Schanz, Schriften aus dem Institut für Forstökonomie der Universität Freiburg 1996, 4 ISBN 3-980 36 97-3-0, pp. 131.

[3]   U. Grober, Veröffentlichung der Bibliothek "Georgius Agricola" der TU Berga-kademie Freiberg 2000, 135 ISBN 3-86012-115-4, pp. 6.

[4]   K. Hasel, Pareys Studientexte No. 48, Verlag Paul Parey Hamburg, Berlin 1985, ISBN 3-490-03316-7, pp. 258.

[5]   H. C. von Carlowitz (1713). Sylvicultura oeconomica. Anweisung zur wilden Baum-Zucht. Reprint der Ausgabe Leipzig: Braun, 1713. Veröffentlichung der Bibliothek "Georgius Agricola" der TU Bergakademie Freiberg, Nr. 135. Freiberg, 2000, ISBN 3-86012-115-4, pp. 414 + Register.

[6]   J. Radkau, Natur und Macht, Eine Weltgeschichte der Umwelt, C. H. Beck, München, ISBN 3-406-48655 X, 2002, pp. 469.

[7]   Pro Silva (Association of European foresters practising management which follows natural processes). Declaration of Apeldoorn. In: Sustainability the Pro Silva way. Proceedings of the 2nd International Pro Silva Congress, Apeldoorn, The Netherlands, May 29–31, 1997, 7–8.

[8]   R. Mosandl, and B. Felbermeier, Vom Waldbau zum Waldökosystemmanagement. Forstarchiv 2001, 72, pp. 145–151.

[9]   WCED – World Commission on Environment und Development (Eds.), Our Common Future. Oxford University Press, ISBN 0-19-282080-X, 1987, pp. 398.

[10]   SRU – Der Rat von Sachverständigen für Umweltfragen, Umweltgutachten, Für eine neue Vorreiterrolle. Metzler-Poeschel Verlag, Stuttgart, 2002, pp. 549.

[11]   E. Williams, P. Charleson, N. Deasley, V. Kind, C. MacLeod, S. Mathieson, E. McRoy, Can environmental regulation ever be sustainable? ERP Environment (Ed.), 9th International Sustainable Development Research Conference, Nottingham/UK, 24.–25. March 2003, Proceedings ISBN 1 872 677 44 4, attachment, 2003.

[12]   G. Weber-Blaschke, G. Painter, T. Molnar, N. Michos, M. Faulstich, Indicator Systems for Sustainable Development in Germany, Hungary, Greece, and the United States – A Comparison. ERP Environment (Ed.), 9th International Sustainable Development Research Conference, Nottingham/UK, 24.–25. March 2003, Proceedings ISBN 1 872 677 44 4, 470-482, 2003.

[13]   D. J. Krieger, Ein universalistisches Modell. Die Gestaltung einer interreligiösen Um-weltethik. In: Politische Ökologie No. 48. Ökom Verlag, München, ISSN 0947-5028, 1996, pp. 25–28.

[14]   M. Suda, and R. Scholz,. Nachhaltigkeit – ein gesellschaftlicher Aushandlungsprozess, Das Papier 6A, V32-V36, 1997.

[15]   H.-J. Otto, Waldökologie. Verlag Eugen Ulmer, Stuttgart, UTB ISBN 3-8252-8077-2, 1994, pp. 391.

[16]   P. Burschel, Das Konzept der Nachhaltigkeit und die forstliche Produktion. AFZ/Der Wald, No. 11, 2003, pp. 566–568.

[17]   P. Burschel, and J. Huss, Grundriß des Waldbaus, Ein Leitfaden für Studium und Praxis, Pareys Studientexte No. 49, Verlag Paul Parey Hamburg, Berlin, ISBN 3-490-00916-9, 1987.

[18]   J. Korhonen, M. Wihersaari, I. Savolainen, Ecological Economics 2001, 39, 145–161.

[19]   J. Korhonen, Journal of Environmental Planning and Management 2002, 45 (1), 39–57.

[20]   J. Korhonen, J.-P. Snäkin, K. Kullki, Analysing the evolution of industrial ecosystems: concepts and application. ERP Environment (Ed.), 9th International Sustainable Development Research Conference, Nottingham/UK, 24–25 March 2003, Proceedings ISBN 1 872 677 44 4, attachment, 2003.

[21]   G. Weber, and M. Faulstich, First Balance of Material Flows for the State of Bavaria (Germany). ERP Environment (Ed.), 6th International Sustainable Development Research Conference, 13th – 14th April 2000 in Leeds/UK, Proceedings ISBN 1 872677 30 4, 2000, pp. 410–417.

[22]   G. Weber-Blaschke, and M. Faulstich, in Müll-Handbuch, Kennzeichen 1406 (Eds. G. Hösel, B. Bilitewski, W. Schenkel, H. Schnurer), Erich Schmidt Verlag Berlin, 2003, pp. 1–34.

[23]   G. Weber-Blaschke, B. Zauner, M. Faulstich, Analyse und Prognose der mineralischen Roh- und Baustoffströme in Bayern, Wasser & Boden 54, Jg., Nr. 1+2, 2002, pp. 27–32.

[24]   H. Cotta, Anweisung zum Waldbau. Arnoldische Buchhandlung, Dresden und Leipzig, 5 verb, Aufl., 1835, pp. 394 + Anhang.

[25]   M. H. Huesemann, Clean Techn. Environ., Policy 5, 2003, 21–34.

2 Sustainable Development: Exploring the Cross-Cultural Dimension

Ortwin Renn

University of Stuttgart, Department of Sociology II, Seidenstr. 36, 70174 Stuttgart, Germany, ortwin.renn@soz.uni-stuttgart.de

Preface

A similar debate about sustainability to the one that we encounter all over the world at present times was conducted almost 45 million years ago. This ancient debate is described in the cartoon below: What you can see is an illustration of a congress of dinosaurs. The dinosaurs are concerned and worried, since their climate is changing and their fate is at risk. The leading dino explains: "The picture is very bleak, gentleman. The world's climates are changing, the mammals are taking over, and we all have a brain about the size of a walnut." The dino was right at his time. All dinosaurs became extinct soon afterwards. We hope, however, that we can mobilize our brains to cope with a global threat that is not much less severe than the ones the dinos faced in their time. The recipe for fuelling this hope is called sustainable development. In this article we will explain the concept and show how the idea of sustainability can be used as a concept for intercultural policy making.

"The picture's pretty bleak, gentlemen ... The world's climates are changing, the mammals are taking over, and we all have a brain about the size of a walnut."

Global Sustainability. Edited by P. A. Wilderer, E. D. Schroeder, H. Kopp
Copyright © 2005 WILEY-VCH Verlag GmbH & Co. KGaA, Weinheim
ISBN: 3-527-31236-6

2.1   Introduction

The concept of sustainable development has enjoyed unprecedented popularity in recent years. It was originally introduced as a microeconomic concept in forestry meaning a strategy aimed at providing wood continually without wiping out the forest [1]. Since the mid-Eighties, though, it has become a popular catchword for attempts to link economic development with maintenance of an ecologically determined carrying capacity. At the 1992 United Nations Conference on Environment and Development (UNCED) in Rio de Janeiro, the concept of sustainable development played a central role. More than 300 pages of recommendations in reference to this concept were collected and published as Agenda 21 [2]. In addition to the United Nations, many other national and international organizations have jumped on the sustainability bandwagon. These include the World Bank, ecological research institutes and corporate groups as well. Entire new research institutes have been set up throughout the world with "sustainability" as their field of concentration.

How can such sudden popularity of a concept taken from forestry be explained? For one thing, it reflects the environmental debate that is raging currently. Public opinion demands that environmental concerns are included in industrial and economic policies. Secondly, it suggests that there is a way to reconcile economic expansion with ecological limits. Hardly anyone doubts that there are limits to quantitative/expansionary growth. The radius r of Planet Earth is a finite quantity, and so is the earth's surface area. Pollution and resource consumption cannot grow indefinitely. The productivity of the environment simply can no longer be boosted as one might like it to be.

The very phrase "can no longer be boosted" hints that productivity, and with it the environment's carrying capacity for human purposes, has been influenced in the past, and perhaps could be influenced again, by human inventiveness. In spite of the fact that an absolute limit is placed on the environment's carrying capacity by ecological conditions, it is the economic and social conditions that determine the productivity of the carrying capacity for human purposes below the absolute threshold [3]. 12,000 years ago, about 5 million people were living on earth, and the carrying capacity – given the productivity attainable at the time, an era of hunters and gatherers – was also limited. Even the agrarian/preindustrial culture that predominated around 1750 had a strictly limited carrying capacity of about 750 million people in the entire world.

Today the world supports almost 6 billion people – with the total still rising. Every nine months, another 80 million people are added – as many as in Reunified Germany. That is a carrying capacity a thousand times larger than that of the New Stone Age, a level still improving with each new gain in productivity. This enormous achievement of human culture has been accomplished with the aid of the five "Promethean innovations": harnessing fire, inventing agriculture, transforming heat from fossil fuels into mechanical energy, industrial production, and the substitution of information for material [4].

However, there are more and more signs telling us that the growth of humanity and rising consumption in those countries in which the per-capita consumption of goods and services is increasing can no longer keep pace with the need for global carrying capacity, despite the accelerated innovation intended to adapt productivity to the development of humanity. The clock of exploitation is running ahead of the clock of natural rejuvenation and human advances in boosting productivity. Today's population density in conjunction with the level of consumption of the rich are claiming a complete transformation of nature into productive environment. At this rate it appears necessary to resort to unique and very limited fossil energy reserves and raw materials, and to exploit regenerative resources until they will no longer be able to regenerate themselves. Humanity depends on nonrenewable energy sources for over 80 per cent of its energy needs. But these will be depleted soon – perhaps not quite as soon as many thought a few years ago – but within a foreseeable time span of decades or centuries at the latest [5].

Even if the speed of innovation cannot keep up with the growth of humanity's desire to get more out of its environment, a national economy's innovative potential is still important. It's too late to go back; the situation will no longer allow it. Without technical progress and without boosting micro- and macro-economic efficiency, things would be much worse [6]. Only with the aid of structural change triggered by innovation can the necessary substitution processes be set off; processes that are needed to bring about innovations in products and production techniques, in order to secure the level of affluence we've achieved. At the same time, structural change means a chance to increase the efficiency of environmental utilization, and thus also the world's carrying capacity for human purposes, as can be seen from past experiences. Therefore, a solution to the environmental crisis of our time must begin on both ends: first, by means of innovation, we must try to improve the productivity of the environment, and thus the earth's carrying capacity. Secondly, we must reduce the demand for consumption of the environment. Both of them are needed: ongoing development of economic productivity in the direction of more efficient use of the available production factors, and of the environment factor in particular, while at the same time adapting demand structure to the ecological conditions governing carrying capacity.

In developing countries, the need to get economic development in line with maintenance of carrying capacity is even more pronounced than it is in the industrialized world. Rapid population growth, the special conditions under which the natural environment is often used there, an unfair distribution of wealth and a productivity level lower than that in industrialized countries are a few of the factors causing many groups in the population of these countries to suffer, as all of us claim to deplore. To ask them to do with even less would be perverse. All the same, it is in these very countries that one can already see exploitation of the environment getting worse, leading to soil erosion, steppification, water pollution, and an even less efficient utilization of fossil energy resources than hitherto.

The term "sustainable development" is a prophetic combination of two words which unites both aspects – economic progress and quality of life – in one vision. This vision of an economic structure that meets all needs of this generation without restricting the needs of future generations is highly attractive, because it reconciles the terms "economy" "social stability" and "ecology" – terms seen so often as opposites – and postulates a generally acceptable distribution rule among generations. Sustainability is a vision of a society living below or near its carrying capacity, but which is nonetheless able to satisfy its economic, social and cultural needs. As attractive as the concept may be, of course, few concern themselves with the question as to how the magic formula can be transformed into reality, leaving it instead to the user to decide which of its two components he really likes best: ecology or economy.

The problem with a concept that so many people share as a common goal, and that has become a political catchword showing up in many discussions without anyone's considering what it really means, is that it can be defined and operationalized however anyone wants. When environmental activists and conservative executives use the same term for their very different purposes, they are sure to be using it with different meanings. Sustainability is used all too often as a patronizing phrase to legitimise one's own interests and to hide concealed conflicts. Then the concept loses it normative effect, which could otherwise be an integrating one ultimately.

Should we then do without the term sustainability? That would be throwing the child out with the bath-water. On the one hand, the concept has to be understood correctly; only then can it serve as a realistic and ethical guide for the future development of both industrialized and developing countries. On the other hand, getting the many players in politics and society to agree on any common ground at all offers a great chance to begin making common strategies to implement a concept. There is already a change underway in society towards more environmentally compatible products, which could be reinforced by strategies emphasizing ecological criteria. The type of structural change called for by the concept of sustainable development always requires that a change be underway in one's own behaviour and own goals. And reflection on those personal aspects involves a vision of the future that is considered worth living for. Sustainable development has become a common vision for many groups and can serve as an integrative engine to keep parties with diverse interests heading for the same goal.

Lest such terms deteriorate into meaningless pleas, sustainable development must be nailed down and laid out into definite requirements. What is needed is a description of the concept that is as exact as possible, and details on how to operationalize it. Such a definition should allow for flexibility – but not arbitrariness – in implementing the concept. At the same time, the concept must have the strength of broadly accepted reasons behind it in order to remain applicable above and beyond the interests of the various players involved.

2.2   Perspectives on Sustainable Development

What does sustainable development mean? Apparently it refers to a type of societal development that makes it possible for future generations to have any essential resources left to use. Such a general answer as this, of course, is not enough to develop an operational strategy. If one has a look at the literature, one finds significant differences in how the term is interpreted and operationalized [7]. The confusion arises mainly from four sources:

- differences in the perspectives and approaches of the various academic disciplines and research traditions concerned with sustainability;
- differences in the values and interests of the players concerned, who perceive the situation and evaluate possible strategies very differently from one another;
- differences in the horizons in which the term is used as regards both space and time;
- differences in the writer's intention in the particular context in which the term is used: whether it is a moral plea, a planning basis or an element of technology assessment.

Although it is hard to classify these differences in meaning systematically, there are two main factors according to our literature review which seem to determine the perspectives of the authors who have written about sustainable development: firstly, the way an author associates specific images with nature, and secondly, the particular home discipline or research tradition from which a writer comes. Concerning the images of nature involved, one can distinguish anthropocentric from biocentric approaches. The former can be subdivided further into utilitarian and protectionistic perspectives. Here is a short rundown of each [8]:

1. Utilitarian perspectives

- Nature as a cornucopia for resource utilization: Nature in this context constitutes a resource base for satisfying human needs. Sustainability means preserving this resource base for future generations.

- *Nature as modelling clay for the creation of cultivated land* (gardens, agriculture, forestry, material cycles): In this context, sustainability means transforming unproductive nature into cultivated land that can bear crops for human consumption and that societies can utilize economically and preserve permanently. Nature is only of use for humankind when transformed into cultivated land. But this transformation depends on natural contingencies and is limited by the need to preserve fundamental, natural material cycles.

2. Protectionistic perspectives

*   *Nature as a wilderness worth preserving:* Nature means the preservation of unspoiled land as a response to the immediate need of humans to enjoy and learn from nature. Preservation is independent of any utilization of the resources to be found there. Sustainability then includes not only the preservation of a resource base, but also acceptance of nature's intrinsic value the way it presents itself without human intervention.

*   *Nature as a fragile object to be protected from human intervention:* Seen this way, sustainability is not so much protecting man's basis for life as it is protecting nature (or the environment that nature has become today) from man's intervention. Any further human intervention in, or more intensive utilization of, the environment is to be avoided.

3. Biocentric perspectives

*   *Nature as the unity of all Creation:* All creatures have, in principle, the right to occupy their own niches in nature. But humans are in a position to expand their natural habitat beyond the extent that nature would normally have planned for them. Therefore, humans have a special responsibility to curtail their demands on the natural environment, so as to enable humans, animals, and plants to coexist close to nature. This context, however, never questions the priority of human interests over those of competitors in cases where conflicts arise for the same resources.

*   *Nature as a place for creatures with equal rights:* All creatures not only have a right to an undisturbed biosphere, they also have the same right as humans to develop their lives within the limits set by the rules of nature. In conflicts over the use of resources, all creatures, in principle, should have the same chances. Only in cases threatening their own lives do humans have priority over the rights of their fellow creatures.

Of course, these five prototypes of authors' images of nature are seldom to be found in their pure forms, and often occur in different mixtures, and in different social and cultural contexts. They make themselves felt in everyday experience as individual preferences regarding the use of natural or quasi-natural biotopes, or as preferences for different schools of environmental policy. It would surely be an interesting empirical challenge to investigate systematically authors' concepts of nature and the environment, and to derive implications from them for environmental planning and policy.

In addition to the images that authors associate with nature, the academic discipline or research tradition to which they adhere plays a major role in their attempts to define the concept of sustainability. Similar to the influence of the images explained above, disciplines too display differences in the meaning of sustainability, although here again there are many hybrid concepts, some of them intentional. These differences come from [9]:

- *Economics:* Sustainability describes an economic system in which future generations will enjoy at least the same level of welfare as the present generation. In the narrow definition of welfare, it can include only marketable, priced goods; the broad definition of welfare preferred by some authors includes non-marketable goods as well such as social and political stability, resilience, and the immaterial conditions needed for subjective well-being. Such a broad definition of welfare could also be termed "quality of life" (both individual and collective). A central tenet in the economic view is that the welfare level is determined by some aggregate measure of individual utility, no matter what elements individuals include in such a utility assessment. Consequently, economists assume that many possibilities exist to substitute artificial capital (such as machines, production processes) for natural capital (resources for production and the environment as a repository for waste). Then sustainability implies a necessity to preserve natural capital only when it is not possible to substitute artificial elements for it. The main evaluative criterion for assessing sustainability is the level of aggregate utility, regardless of whether some of its components are irreplaceable.

- *Ecology:* Sustainability means the use of natural resources to the extent that the carrying and regenerative capacities of the corresponding ecosystem are not jeopardized. The focus is not on individual resources, whether they are production inputs or waste repositories, but rather on the interaction of related resources within one ecosystem. The system may only be disturbed by human intervention as long as the functionality and regenerative capabilities of the system are not jeopardized. Within the field of ecology, there are different methods of measuring the degree of anthropogenic influence on ecosystems. One especially meaningful one is the degree to which net primary production is used for human purposes. Annual net primary production (NPP) is defined as the amount of energy left after subtracting the respiration of primary producers (mostly plants) from the total amount of energy (mostly solar) that is fixed biologically [10].

- *Physics (and other natural sciences):* Sustainability is the ability of biological systems to create permanent order (negentropy) using solar energy. The time scale over which a well-ordered world can be sustained is limited only by the sun's life cycle. This interpretation of sustainability is based on the Second Law of Thermodynamics. As every transformation of energy increases entropy (which is disorder), all physical processes are ultimately aimed at disorder, and thus a standstill. Biological systems, however, have the ability to create order on a limited scale by using solar energy, and to keep the inevitable entropy outside their own systems. The practice of human societies of transforming more energy by consuming fossil fuels than the sun offers on a continuous basis, and of using more and more of the order-rich products created by biotic systems for its own purposes decreases the biological systems' ability to survive and regenerate themselves (analogous to the ecological approach). Therefore the central issue in connection with sustainable development is continual preservation of the biotic systems' potential to create negentropy.

- *Chemistry:* Sustainability means closing anthropogenic material cycles. All resources used by humans (as either production factors or waste repositories) have to be integrated in closed material cycles in such a way that the outputs (wastes) of one player can be used as input for another player (humans, plants, animals) within the limits of the second law of thermodynamics. Waste, for instance, has to be reused as an energy carrier or recycled into new products or services. Non-reusable waste should be in a condition such that it would decompose into non-toxic, non-polluting substances.

- *Social sciences:* Sustainability means the social and cultural compatibility of human intervention in the environment with the images of nature and the environment constructed by different groups within society. It does not matter whether there is a real environmental crisis in the sense of facts proven by natural sciences. The social science perspective takes as its starting point the social perception that there is a crisis. Social perception of environmental conditions is always selective and suggests that certain (culturally determined) assessment patterns are to be used. Statements on sustainability are thus preferences of the current generation as to the levels of environmental quality and quality of life that it allows itself, and those that it will allow future generations. Issues of distributional equity dominate in social visions of sustainability.

These differences among disciplines are not only semantic, but have further implications. The economic understanding of sustainability, for example, places sustainability in the context of scarcity. The goal of sustainable development is to express the relative scarcity of the resource "environment" in comparison to other production factors through eco-taxes (Pigou), or negotiations for environmental rights (Coase) [11]. Given a dynamic price system, the price based on scarcity provides for optimal allocation (reflecting the relative scarcities). The term "relative" not only refers to scarcity in comparison to other production factors, but also to the urgency with which the consumption goods are in demand. Environmental goods are also subject to this exchange in a free market economy; they can be exchanged for other goods. However, most economists conclude, the prices prevailing for most environmental goods are below the actual level of scarcity for reasons of market failure or market imperfections (not further explained here). Therefore, measures are absolutely necessary to correct the price distortion by governmental interventions.

Many approaches to resource economics assume that production factors can be substituted for one another in any proportions. However, this view is irreconcilable with most natural science perspectives. Natural scientists claim that there are absolute limits to the utilization of the environment, which may not be exceeded, no matter how beneficial it might appear from a purely economic viewpoint. This difference in approach also expresses itself in the types of measures that the respective proponents call for: Economists prefer financial incentives as a means of correcting a market imperfection retroactively, because incentives help reach ecological goals in a cost-efficient manner, thus helping conserve both natural and human resources. Natural scientists, on the other hand, usually prefer (preemptive) mandated regulations based on absolute limits, because regulations reflect the absolute limits of carrying capacity. The social scientists, yet again, emphasize the constructive character of the concept "sustainability" and focus on equity conflicts, which can be solved neither by incentives nor by regulations, but demand specific political measures of redistribution or communication (legitimisation).

Despite the variety of approaches to, and perspectives on, sustainability that exist, the particular challenge for this project is to develop a basic definition that is acceptable to most authors and practitioners, and then to integrate the particular strengths of each perspective into it. In order to meet this challenge, two requirements have to be fulfilled:

a) It may be possible to develop a generally valid definition of sustainability, which reflects the views of most representatives of the concept. Eventually, however, such a consensus is likely to break down when we try to operationalize it. Instead, therefore, it seems appropriate to select modules to operationalize the principles for the different types of applications; they support one or the other of the above perspectives depending on the particular context of the problem. For instance, setting absolute limits or standards, as suggested by the physical interpretation of sustainability, may be more appropriate for managing nonrenewable energy resources (such as coal or oil) than the economic considerations regarding relative scarcity. As for the issue of maintaining natural biotopes, a protectionist perspective may be more appropriate than a utilitarian one. In principle, the perspective should be used which appears best suited on the basis of general criteria.

b) Every attempt to operationalize the term sustainability has different implications when we consider global, national or regional levels, and when we apply it to countries with different levels of economic development. Developing such a wide range of modules for many areas of application would go beyond the scope of this article. However, when we look at sustainability from an intercultural perspective, it is essential that we sustain the basic idea of sustainability and, at the same time, provide sufficient flexibility for the different cultures to search for their specific conceptualisation of this goal.

In the following two sections we will describe the requirements for an intercultural concept of sustainability, and then develop from that analysis our concept of sustainability, in which the various perspectives can be integrated.

2.3   Prerequisites for Sustainable Development

Modern societies cannot live in harmony with nature, but rather at best in harmony with their environment. Nature and the environment are two different phenomena, however they share some basic natural cycles that make them both function. Environment is a product of culture – i.e. human-made, not something humans found waiting for them. In transforming nature into environment, human societies invest knowledge, thought and work; individuals and groups express aesthetic needs in the process as well. That yields an ecological added value. This ecological added value has comprised society's basis for existence since the New Stone Age. The transformation from natural to cultivated land is a generic development that can be found in all human societies.

Thus the natural world became first an agricultural, then an urbanized world; natural ecosystems (originally more or less self-regulating) all around the world were transformed into anthropogenic ecosystems. These anthropogenic ecosystems are what yield food and products for billions of people. Modern societies cannot live from nature alone, but they cannot survive without nature either. Humans use about 95% of the earth's tillable soil as their environment: agriculture, forestry, settlements and infrastructure facilities leave little room any more for the original nature [12].

But our future does not depend on maintaining nature in its original state, but rather on maintaining the anthropogenic ecosystems that we call the environment. Having an intact and productive environment is an indispensable condition for human existence and culture. If the environment degenerates, modern society's existential basis disintegrates. The anthropogenic ecosystems from which humanity lives require a continual input of energy (or to be more exact, negentropy) and continual preventive intervention to keep them from breaking down. Nothing in today's world regulates itself automatically for the human benefit; social forces have to intervene continually to correct environmental conditions. That's why knowledge about the ecological basis of our life is so important. Using such knowledge creatively is a prerequisite for the survival of humanity.

Carrying capacity is a concept of central importance for continuing humanity's economic survival [13]. For humans, the carrying capacity of a region is determined by the techniques of economic conversion or production used. These increase carrying capacity, but not beyond certain limits – neither regionally nor globally. There are biological reasons for that: since nobody knows of any technique to boost sufficiently global photosynthesis, and thus net primary production, there are objective limits to carrying capacity. Annual net primary production (NPP) is a finite quantity; it refers to the amount of energy left after subtracting the respiration of primary producers (mostly plants) from the total amount of energy (mostly solar) that is fixed biologically. Everything that is alive lives from NPP. But it's like living from hand to mouth. There simply are no significant reserves of it. Modern societies have already laid claim to – or influences for their own benefit – 40% of all land surfaces' potential net primary production. Starting from that basis, attempts to expand carrying capacity would come up against limits very soon, even if humans were to grant all living things besides himself no more right to live than as their domesticated animals and potted plants. In many regions of the earth, societies have already reached – and in some instances, even exceeded – the critical load, the limit to irreversibly damaging the environment.

In addition, social constraints limit the maximum yield of NPP for productive purposes. A distinction must be made between the earth's biophysical carrying capacity on the one hand, and its social capacity on the other. Even if it is theoretically possible to make much of the remaining 60 per cent of NPP usable for mankind, it would be neither realistic nor desirable, because humans have social needs for environmental quality that goes beyond the environment's economic value. These needs deserve respect in addition to their physical needs. Take the social need for parks as an example – parks in which there is still at least one leaf not earmarked for some economic purpose. Increasing the percentage of NPP used is therefore of limited value for coping with the environmental crisis. Instead further economic development must boost the efficiency of how NPP is being currently utilized and transformed by humans. The real challenge consists of finding a way to set off a development process which would use the potential of efficiency boosts, while still respecting the absolute limit that applies to the yield capacity of the environment for productive purposes.

2.4   Sustainability: Essentials for a Realistic Cross-Cultural Concept

What conclusions can be drawn from these thoughts on carrying capacity and economic development to help better understand "sustainability?" One metaphor commonly used to characterize sustainable development is to see it as an economy in which people can live off the interest generated by capital stock without invading the capital itself. Given our population density today, to think that we have or could accomplish such an economy would be an illusion. If we were to limit the energy supply for the almost 6 billion inhabitants of this earth to annual solar input alone, leaving all non-renewable resources untouched, reverting to agricultural production methods that are purely extensive and taking other measures necessitated by such a limitation, it would bring about a social catastrophe that would make all catastrophes ever known throughout the history of mankind look mild by comparison. Nor could it be legitimised by any argument of intergenerational equity.

The high carrying capacity that modern society has established already requires not only transformation of nature into productive environment; it also implies dipping into nonrenewable reserves of fossil energy and raw materials, while voraciously exploiting renewables. This "strategy" cannot be maintained much longer. The dilemma is clear: our current economic beast is gnawing away at our capital stock of natural resources, thus digging its own grave. Likewise the call for preserving natural capital stock for future generations would compromise the claim of the present generation to enjoy at least a modest quality of life, considering today's population densities. Not even a development oriented around sustainability can avoid living from nature's capital stock. Man can only stretch natural supplies, not make them last forever.

The ultimate goal of sustainable development must be to maintain both the productivity of nature and the environment, as well as immaterial gains in their utility, for as long as possible. Given today's economic structure, this goal is still far off. Neither in the area of energy utilization nor in the consumption of nonrenewable resources is it imaginable that we continue with today's production techniques and rates of utilization over the long term. But today it is possible, and would be sensible, to develop ways and strategies of coming closer to this goal. At this point in human history, we need to focus on the path to sustainability rather than to implement a sustainable economy at once. This path towards sustainable development is characterized by the following four principles [14]:

- *Increasing resource productivity:* The utility gained from the use of natural resources must be improved continually, so that the natural assets consumed to produce a given quantity of goods and services sink continually. That requires growth in the stock of artificial capital as the only way of keeping the welfare of a country at least constant.

- *Acknowledgment of the limits of substitution between natural and artificial capital:* In traditional economic theory, the monetary value of any good determines the rate at which it can be exchanged for other goods. For certain goods, however, this rule of exchange does not work, as production or consumption of them incurs such high external costs that infinite compensation of those damaged would be needed to make the transaction economically efficient. If, for instance, the quality of the air we breathe were jeopardized by emission of a toxic gas, there would be no macroeconomic benefit great enough to compensate humanity as a whole for the consequences of the poisoning. Within a few minutes, there would be no humanity to compensate anymore. For this reason, the natural cycles necessary for mankind's survival must be identified (by ecological science primarily), and protected by political measures.

- *Focus on the resilience of anthropogenic ecosystems:* Renewable resources would appear to be utilizable for the long term, since they regenerate themselves. Their regeneration capacity lasts, however, only as long as they remain invulnerable to changes in their natural and anthropogenic environments. Monocultures, or ecosystems intensified so as to produce a maximum yield, can only yield a constant output per unit of time under optimal internal and external conditions. However, if those conditions change (by having had excessive strain placed continuously on a production medium such as the soil, or by interaction with the environment such as being attacked by diseases and parasites), these systems break down, i.e. exhibit an abrupt decrease in yield – or worse yet – a permanent reduction in, or disappearance altogether of, their yield potential.

- *Incorporation of social and cultural values in man's relationship to the environment and nature:* Whereas economists emphasize the utility of the environment and nature that can be accounted for in prices, and natural scientists concentrate on the function of cycles and structures vital to the system, social scientists attach a number of aesthetic and symbolic values to the environment and nature, which are of central importance to individual and social well-being. They are however, undervalued in conventional economic processes because they belong to no-one but the public; and public resources are either neglected or overexploited in decentrally organized economies (the so-called "tragedy of the commons"). This is the reason that when we speak of the "quality of life", we are referring to a combination of utility and the attachment of social and cultural values to certain living conditions delineated from the enjoyment of environmental quality. The degree to which these cultural values play a role in the non-economic evaluation of natural and artificial goods depends on the basic cultural traditions, religious beliefs and social symbolizations Both the concrete utility that a society derives from the use of the environment, and the vitality or satisfaction that different cultures derive from experiencing, or even communing with, the environment and nature must be taken as a yardstick of sustainability.

For this reason, we would propose the following definition for the term sustainable development. Sustainable development denotes a process by which the capital assets of natural, social and cultural resources are preserved to the extent that the quality of life available to future generations will not be inferior to the quality of life of the present generation. This definition points the way to the goal of utilizing the resources of human societies for the long term. It is a matter of ensuring continuity in the supply of natural and societal resources. This implies three requirements:

- First, a sustainable path aims at preserving the biosphere's reception capacity for material flows that have been anthropogenically caused or influenced by human activities. This must be done so that future generations as well can enjoy a level of welfare similar to that of today's generation.
- Second, a sustainable path aims at preserving the capacity for technological innovation, economic welfare and social institutions. This classic form of capital includes the infrastructure for economic activities, the functioning of economic and political transactions, and the preservation of social institutions that govern production, education, political stability, and social coherence.
- Third, a sustainable path aims at sustaining cultural traditions, beliefs, values and fundamental convictions that constitute individual and collective identity within limits of universal human rights and the necessities of a global economy.

Our definition deviates from the most common three-component definition of sustainability (ecological, social and economic component) in two major ways [15]: We have integrated the social and economic component into one single aspect of sustainability. Economic affairs are always social in nature, and social aspects that defy economic consequences are worthless for the analysis. In the terms of modern economics, social consequences are part of economic welfare as much as the monetary equivalent of goods and services. For the third component, we have included as a separate entity cultural beliefs and values [16]. This broadening of the traditional concept is justified for several reasons:

- Cultural beliefs and values are prime motivational agents for individual and collective actions; they co-determine (together with interests and incentives) the way that human act in favour or against sustainability [17];
- If preservation of certain goods is not embedded in the portfolio of what a culture prescribes as valuable, any attempt to enforce this preservation will be futile in the long run even if force is being used;

- Culture is a good in itself that needs to be preserved as the main component of personal and collective identity. It provides "purpose" to life and is thus essential for human dignity. As much as we need to preserve the natural and social capital, we are in urgent need of the cultural capital that provides the basic motivation and ontological security for human beings.

Although the definition contains economic, social, cultural and ecological concepts, the call for sustainable development is an external norm arising out of an ethical motivation, which does not result automatically from any efficiency criterion of economics, or any concept of limitations in ecology. Instead its justification is to be sought in the principle of intergenerational equity. Why should our descendants be worse off than we are? Sustainability is a constraint on economic development much like human rights; it does not question – much less cancel out – the basic intention of economics, namely to orient human economic activity towards maximum efficiency nor does it ignore the absolute limits imposed by the carrying capacity for human purposes. It provides a positive vision for economic and cultural development taking into account the vulnerabilities of human societies and the innovation potential for improvements.

Sustainability implies an integration of economic efficiency, essential resource preservation and the continuation of social and cultural identity. There might be conflicts between these three objectives. For instance, there might be ethical reasons for protecting certain animals or for keeping animals in more humane living conditions which do not yield a monetary benefit, but can be justified out of a respect for the Creation. Biodiversity can also be justified intrinsically (by consciously placing a value on it), and is not dependent on a line of reasoning that anticipates a possible future need as part of a consideration of monetary benefits (which is usually not very convincing anyway). For such a divergence from the cost/benefit principle to be rational, however, decision-makers have to be aware of the costs of the measure before they consent to it. Granting ethical reasons validity and opting for a solution other than that of maximum utility can always be legitimated vis-a-vis future generations, even if the future generations would not be able to follow the ethical reasoning that lead the present generation to do so [18]. The reason that our (European) ancestors did not want to eat horsemeat surely does no harm to us, even if we only abstain from horsemeat for traditional reasons. The deliberate decision not to use a resource for economic purposes can never compromise the possibility of future generations to improve their level of welfare.

To express the desired integration of ecology, socio-economics and culture as a basic concept of sustainable development we have taken refuge in the term quality of life. This term is useful for expressing the association between utility and the attachment of immaterial value. Furthermore, it comprises an abstract idea that will remain independent of transient fashion. Quality of life includes the social and cultural experiences that determine personal well-being, and the objective conditions that make such experience possible. Even if it is impossible to know exactly what requirements future generations will place on their social and cultural experience, we can determine preemptorily what potential will have to remain to give these generations the chance of having certain experiences (which are desirable ones today). If they don't want to use that potential, no problem; it would not even be a waste of resources because generations after them might change the trend again. However, if coming generations are deprived of the chance to have these experiences, they lose quality of life. Basically the project assumes that the physical, economic, cultural and social foundations and conditions associated with a certain level of quality of life will have to have some significance for some future generation as well – regardless of whether the more immediate descendants need, or even appreciate, them.

It would be an illusion to assume that the present state of society, economics and environment needs to be sustained for generations to come. Firstly, nature itself acts an agent of change and influences the development of environmental transformations over time. Secondly, human forces will always alter the environment through economic, social and cultural activities. What needs to be preserved are those elements of the natural, social and cultural capital that individuals of the present generation value for economic, social, or cultural reasons. Sustaining quality of life means therefore preserving those elements that humans appreciate today as options for future generations of which we do not want to deprive them. If they don't value these conditions, their quality of life would not be reduced. If they value aspects of the environment that present societies have not included in their portfolio of valuable assets, it is our hope that these losses can be compensated by the increase in artificial capital that we accumulate and hand over to the next generation. Any other solution to the problem of what to select for preservation would either overtax the capabilities of the present generation to predict future needs or increase the probability that future generations would have less quality of life than today's generation. How to draw the line between the essential elements that we need to preserve and those that can be further transformed and changed cannot be determined in advance. This is a task of continuous communication and negotiation within a culture and between cultures.

2.5 Qualitative Growth as a Prerequisite for Sustainable Development

The special attraction of the term "sustainable development", as already mentioned, lies in the way that it unites two apparently contradictory demands: one of them for non-destructive utilization of the natural, social and cultural capital and the other for further economic development. A number of authors have repeatedly emphasized that the term "development" should cover only structural change, but not growth in the economic sense. In our considerations, we do not want to dismiss the idea that a zero-growth economy is possible in theory, and it certainly appears to be more in harmony with the concept of sustainable development than a growth-oriented economic order [19]. However, we must continue to regard this solution as problematical for the following four reasons:

1. As long as people associate consuming ever more products with welfare, the strategy of relying less on one's natural capital stock can let the level of welfare remain constant, or let it even improve, only if the assets of artificial capital grow at the same time. In principle, taking strain off the environment by using it more efficiently is only possible by boosting artificial capital, unless welfare is allowed to drop. There are many indications that there may be no immanent limits to boosting artificial capital. This hope is supported primarily by the realization that "knowledge" is a production factor with special characteristics: it displays no evidence of wear whatsoever, i.e. knowledge, once acquired, can be reproduced as often as desired (no rivalry arises). Nor does one presumably have to expect diminishing returns in the production of knowledge.

2. After the fall of Communism, various forms of free market orders have become the new world standard. Within these economic orders, structural change takes place only because there is hope for growth. In principle, growing and shrinking industries could balance one another off; but no one can tell in advance. As long as one maintains the freedom to invest, without which free market systems would not be feasible, market players have to be allowed the hope of growth.

3. If one assumes that the prices in a free market economy reflect relative scarcities, then there is really no argument why a condition in which everyone feels better than before should be rejected. As long as the market imperfections that have caused natural capital to be underpriced can be compensated for by establishing new exchange ratios, there is no reason why people shouldn't be allowed to try to improve their individual benefits. If we were able to internalise the external costs of using natural assets as well as all other social costs, then economic growth would reflect the improvement of human welfare. There may be limits for internalising external costs which would necessitate direct governmental interventions. Such limits, however, do not compromise the function of economic growth for stabilizing and even improving economic welfare as long as external costs are accounted for in some way. The argument often put forth in the discussion of growth to the effect that growth enlarges the gap between rich and poor is not aimed at growth per se, but rather against the unfair ratio according to which any surplus wealth attained is distributed.

4. Zero-sum growth might make sense for industrialized countries living in relative affluence, but not for the people in developing nations living in poverty and desperation. A concept such as sustainable development should apply to all, or at least its basic principles should, even if individual elements have to be adapted from region to region. The fact that population growth in developing countries comprises a particularly critical factor does not contradict the basic tenets developed here (refer to the third phase of qualitative growth below).

For these four reasons, it appears appropriate to acknowledge the mechanism of growth as an integral component of a sustainable economic system, although it has undoubtedly been one of the causes that have had negative effects on the environment and indigenous cultures in the past. But we intend to fill it with new life in such a way that it would no longer be in contradiction to the second prerequisite, the one calling for maintenance of the natural basis of life, for which there is no substitute. The term "qualitative growth" has come into use meaning economic development that is controlled or influenced according to certain external criteria [20].

Qualitative growth in our context means a process during which resource productivity increases continuously with economic welfare. The increase in the performance of a national economy attained by growth has to be continued, but by exploiting less of natural, social and cultural resources. Every unit of the three major capital assets (natural, social and cultural) should become more productive. It is difficult to determine what qualitative growth means in terms of the social and cultural capital (both do not constitute zero-sum-games, but can easily be exploited or depleted). But one can explain the concept well by looking at the natural resources. The objective would be to create a second event parallel to the historic achievement that we have seen, during which an enormous increase in the productivity of work per hour was attained. Now we need to initiate a new era characterized by a rising productivity of natural resources (per unit of energy or raw materials). Qualitative growth in this respect is thus characterized by a further increase in productive services although the use of resources and environmental damage decrease. This becomes possible because knowledge and immaterial services are substituted for material resources and manual work: structured knowledge and software replace raw materials, exergy and time. One can distinguish among three stages of qualitative growth [21]:

- In Stage I, qualitative growth means a continuous decrease in resource use per unit of domestic national product [22]. Every product has to use less resources than the one replacing it. This also applies to utilization of the environment as a repository for no longer recyclable waste. Most industrialized countries have reached this first phase of qualitative growth for most economic goods.

- In Stage II, qualitative growth means a continuous decrease in resource use per capita. Here we have the additional criterion that the saving effects derived from better use of the environment have to be greater than any additional use of resources resulting from an increase in production or consumption. Only such sectors would grow in Stage II that bear promise of creating an disproportionately large value added while using less of the environment. This second stage of qualitative growth has been reached in the manufacturing sector for only a few products thus far.

- In Stage III, qualitative growth means a continuous decrease in resource use per national economy, and thus indirectly on the global level as well. Stages II and III are identical for societies without population growth. In this third stage, which applies specifically to countries in which a high birth rate or migration has caused the population to grow, the absolute utilization of resources also has to decrease. Then economic structural changes not only have to compensate for the greater consumption demands of each individual, but also for the collective demands caused by population growth as well. This third stage of qualitative growth therefore will be the hardest one to accomplish. Success will only be possible if measures for checking population growth are enforced in parallel to structural changes.

Qualitative growth is not an illusion. Creating added value using software is routine; the history of technology is full of examples of using innovation to find substitutes for scarce goods. The new dimension of substituting software and knowledge for material and energy that has been created through the progress of science opens a new dimension of qualitative growth. Such innovations lay the foundation for the requirements needed to bring about Stage II of qualitative growth encompassing all sectors. Of course, even future technologies cannot provide added value for nothing. Economic growth decoupled from resource consumption is neither free of side effects, nor can it reduce the input of raw material and energy indefinitely. No economic structure can guarantee a one-hundred-percent closed material cycle – at least not given today's population density. But there is a lot of room for improvement over business as usual.

2.6   Outlook

More than ten years after the Rio Declaration with its emphasis on sustainability it is time to reflect on the experiences and insights gained over last decade. One can detect a positive and a negative trend: the positive experience has been that almost all nations and almost all relevant social groups have approved this concept and pursue, at least verbally, a path towards sustainability. The flip side of this accomplishment is the enlargement of the definition of sustainability. Any practice or any activity that special interest groups want to preserve are declared as cases for sustainable development. When sustainable profits or sustainable social services are cited as examples for sustainability in the political discussion, the term becomes useless since it does not exclude anything.

At the same time, however, the term has suffered under a severe cultural bias. It has been coined and developed within the dominant western culture and has been inspired by the efficiency criterion of market economies. Both limitations may be legitimate for a discussion within the OECD countries but they pose limits to the necessary expansion of the concept to other cultures and other disciplinary perspectives. Our approach to a cross-cultural conceptualisation of the term is both a restriction and an enlargement of the original idea:

- the term includes three major components, the natural, socio-economic and cultural capital that needs careful consideration about what needs to be preserved and what needs to be changed;
- the term includes a requirement for a continuous dialogue as to specify what needs to be preserved among the three components and what needs to be changed. The meta-criteria for this selection can be derived from the rationales of the three components:

sustaining natural resource productivity; sustaining social and economic welfare and justice; and sustaining individual and collective identity.

These meta-criteria are general orientation marks for a more detailed delineation of concrete criteria and requirements. This delineation cannot be prescribed by scientific reasoning but depends on a discursive process. To identify public values and integrate facts and values into a joint decision making effort, a communication process is needed that build upon intensive dialogue and mutual social learning. Without consulting pubic interest groups and those who are affected by the application of the criteria selected for generating and evaluating sustainable paths, a meaningful synthesis of expertise and public concerns cannot be accomplished. A recent report by the National Academy of Sciences calls for an integration of assessment and discourse forming an "analytic-deliberative" approach [23]. The objective is to design cooperative planning processes in which uncertain outcomes are discussed with representatives of the affected public and the evaluation of options is performed in an active dialogue between experts, stakeholders, and members of the general public [24]. Dialogue and negotiations are even more essential in cross-cultural settings. These setting demand processes for creating mutual understanding, a common problem definition and a joint effort for selecting criteria and indicators for all three components of sustainability [25].

In the long run, cultural values and commitments will be the decisive factor for the implementation of sustainability. Focusing on natural and economic resources will limit the view of the development potentials and impede successful strategies that are in line with the convictions of those who need to take actions. If we are able to integrate the three components of sustainability as outlined above there is a realistic chance that this concept will slowly diffuse into the majority of cultures and become a universal yardstick for judging societal progress in the time to come.

References

[1]   W. Peters, PhD thesis, University of Hamburg (Germany), 1984.
[2]   V. Hauff (ed.) in Unsere gemeinsame Zukunft. Der Bericht der Weltkommission für Umwelt und Entwicklung (Brundtland-Bericht), Greven 1987, p. 46 as well as Agenda 21: Programme of Action for Sustainable Development, United Nations Press: New York, 1993.
[3]   W. E. Rees in The Ecologist, Vol. 20, 1990, p. 18–23; cf. S. Postel in L. Brown (ed.), State of the.
[4]   O. Renn and R. Goble, Sustainable Development 1996, 6, 34–56.
[5]   H. Mohr, Qualitatives Wachstum. Stuttgart: Eggbrecht 1995.
[6]   D. Cansier, Nachhaltige Umweltnutzung als neues Leitbild der Umweltpolitik. Discussion Paper No. 41, Economics Department of the University of Tubingen. Tübingen (February 1995), p. 6 ff.

[7]   A variety of definitions can be found in: R. Allen: How to Save the World, London: Kogan Page, 1980; L. Brown, C. Flavin, S. Postel (eds.): Saving the Planet, New York: Norton 1991; R. Goodland, G. Leddec: Neoclassical Economics and Principles of Sustainable Development, Ecological Modelling, 38 (1987); D. Pearce, A. Markandya, E. B. Barbier: Blueprint For a Green Economy, London: Earthscan 1993; R. Sollow: An Almost Practical Step to Sustainablility, Washington: Resource for the Future, Washington 1992; World Commission on Environment and Development (WCED): Our Common Future, Oxford: Oxford University Press 1987; Mannheimer Declaration, prepared by the Center of Technology Assessment in Baden-Württemberg, Manuscript. Stuttgart 2002.

[8]   The classification is taken from: A. Knaus and O. Renn: Den Gipfel vor Augen. Unterwegs in eine nachhaltige Zukunft. Marburg: Metropolis 1998. See also: M. Thomson: "Socially Viable Ideas of Nature: a Cultural Hypothesis", in E. Baarknand U. Svedin (eds.): Man, Nature, Technology, London: Macmillan Press 1988, pp. 57–79.

[9]   Classification taken from O. Renn: A Regional Concept of Qualitative Growth and Sustainability – A Pilot Project for the German State of Baden-Württemberg. Working Report No. 2. Stuttgart: Center of Technology Assessment in Baden-Württemberg, April 1994.

[10]   This definition is taken from Peter M. Vitousek, Paul R. Ehrlich, Anne H. Ehrlich and Pamela A. Matson, Human Appropriation of the Products of Photosynthesis, BioScience, Vol. 36, No. 6 (June 1986), pp. 368–373.

[11]   R. Solow: The Economics of Resources or the Resources of Economics, in: American Economic Review, Vol. 64 (2), 1974, pp. 1–14.

[12]   H. Mohr: Vom quantitativen zum qualitativen Wachstum, Informationsdienst Soziale Marktwirtschaft, Vol. 35, 1993, pp. 1–6.

[13]   cf. Mohr 1993; see also W. R. Catton.: Overshoot: The Ecological Basis of Revolutionary Change. Urbana: Illinois University Press 1980.

[14]   O. Renn,.; R. Goble and H. Kastenholz: How to Apply the Concept of Sustainability to a Region, Technological Forecasting and Social Change, Vol. 58, 1998, 63–81.

[15]   cf. J. Huber: Nachhaltige Entwicklung. Strategien für eine ökologische und soziale Erdpolitik. Berlin: Edition Sigma 1995.

[16]   Similar approaches in: R. van der Wurff: Sustainable Development a Cultural Approach – A Report for the Forecasting and Assessment in Science and Technology (FAST) Programme, Brussels: Commission of the European Communities, 1992; I. J. Simmons: Interpreting Nature. Cultural Constructions of the Environment, London: Routlege 1994.

[17]   P. MacNaghten, and M. Jacobs in Global Environmental Change, Vol. 7(1), 1997, pp. 5–24.

[18]   W. Korff: Umweltethik, in: M. Junkernheinrich, P. Klemmer and G. R. Wagner (eds.): Handbuch zur Umweltökonomie, Berlin: Analytica 1995, pp. 278–284.

[19]   See the pros and cons to economic growth in: H. E. Daly and K. N. Townsend (eds.): Valuing the Earth, Cambridge: MIT Press 1993, pp. 267–274; R. Kappel. Von der Ökologie der Mittel zur Ökologie der Ziele? Die Natur in der neoklassischen Ökonomie und ökologischen Ökonomik, in: PERIPHERIE, No. 54 (1994), pp. 58–78 as well as. D. Pearce, Blueprint 3: Measuring Sustainable Development London: Earthscan 1994.

[20]   Majer (ed.): *Qualitatives Wachstum*, Frankfurt/Main: Campus 1984.

[21]   O. Renn and R. Goble: A Regional Concept of Qualitative Growth and Sustainability: Underpinnings for a Case Study in the State of Baden-Württemberg. Sustainable Development, Vol. 6 (December 1996), 34–56.

[22]   As long as we don't have a better and commonly accepted indicator for measuring overall welfare, we will stick to the conventional (and in many aspects problematic) economic indicators such as the gross domestic product.

[23]   P. C. Stern and V. Fineberg: Understanding Risk: Informing Decisions in a Democratic Society. National Research Council, Committee on Risk Characterization, Washington: National Academy Press 1996.

[24]   D. Fiorino: Citizen Participation and Environmental Risk: A Survey of Institutional Mechanisms, in: Science, Technology, & Human Values, Vol. 15, No. 2 (1990), 226–243.

[25]   O. Renn and A. Klinke: Public Participation Across Borders, in: J. Linnerooth-Bayer, R. E. Löfstedt and G. Sjöstedt (eds.): Transboundary Risk Management. London: Earthscan 2001, pp. 245–278.

3   Sustainable Development and Cultural Diversity

Hartmann Liebetruth

Bergische University of Wuppertal, Department of Print & Media Technology,
Rainer-Gruenter-Str. 21, 42119 Wuppertal, Germany,
liebetruth@kommtech.uni-wuppertal.de

3.1   Introduction

The term of "sustainability" was introduced into the political discussion after the historical conference of the United Nations in Rio de Janeiro in 1992. The objective of this
conference was to come to agreements between the member states of the UN on measures and actions to be taken for the protection of the environment and for saving the
natural resources to the highest possible degree. Water and air must be protected from
pollution, soils from exhaustion, the climate from man made changes. These measures
are considered necessary to be taken for the sake of future generations, i.e. for not putting at risk the fundamentals of life of future generations by whatever activities of the
present one. This request has been put by the environmentalists in the words: "Our generations has not taken property of the earth, it has only borrowed it for use". Consequently each generation must pass the earth to the subsequent one in the same condition
that it was received.

There is no doubt that most harm to the environment results from certain economic
activities aimed at mostly short term business success. Thus economic and ecological
aims are often in a conflict with each other. Sustainable development is therefore defined
as "meeting the needs of the present generation without compromising the ability of the
future generations to meet their needs" (Brundtland Commission Report, 1987). Business decisions, therefore, have to include ecological considerations for protecting the
environment and hereby ultimately ensuring long-term business success.

There is also general agreement on the fact that sustainable development requires
stable social conditions. Fair distribution of a nation's income and its wealth, the existence of social welfare systems, equal opportunities for education, protection of minori-
ties' rights and similar must therefore included at political and economical decision
making. At first sight, social development appears to be contrary to economic aims.
However, it is also obvious that a prosperous economy needs a stable society. "Sustain-

Global Sustainability. Edited by P. A. Wilderer, E. D. Schroeder, H. Kopp
Copyright © 2005 WILEY-VCH Verlag GmbH & Co. KGaA, Weinheim
ISBN: 3-527-31236-6

ability therefore is about ensuring long-term business success while contributing towards economic and social development, a healthy environment and a stable society". The need for decision makers in politics and economy therefore is to address these three broad components, which are sometimes described as people, planet, profit. This concept is known as the so called triple bottom line (TBL), which means a set of goals which are to some extent contradictory to each other but at the same time necessary conditions for achieving any success in other fields of action. It is quite similar to decision making in the field of economics, where a set of targets is also addressed at the same time. Economic growth, full employment, stability of prices, social welfare, balance of the national budget as well as of foreign trade, environmental protection and the like are economic objectives which can be each a sub goal in relation to any other one, but they can also block any progression in any of these fields of action.

In theory on the level of micro economics sustainable development can be dealt with in many ways:

1. Defining a function to be maximised, as for example a business's long term profits or its shareholder value subject to defined restrictions which refer to certain levels of the environmental health and social stability. For example the emission of carbon dioxide must not exceed a given amount or medical service is provided to everyone without regard to income. These restrictions may be prescribed by governmental rules or set by the firm's own policy (corporate governance) or both.

2. The mechanisms of cost accounting are amended in such a way that they reflect the full cost of a produced good or service, i.e. they must include the so called externalities in addition to the customary factors. Externalities reflect the cost of consuming natural resources (depletion), cost of eliminating damages resulting from its use, cost of waste disposal, the cost of social security etc. It is a matter of course that such full cost accounting results in higher cost of the finished goods and commodities. However, such increases do not reflect additional cost, but shifts from the anonymous public – represented by the government – to those who are at its origin and who have therefore caused them.

3. A combination of both mechanisms.

Whatever way of dealing sustainable development is chosen, the objective clearly is to support all efforts aimed at saving such cost and hereby making substantial contributions to protect the environment to the highest possible degree. It is quite obvious that the measures taken for protecting the environment must be applied world wide in order to avoid competitive disadvantages to those who are submitted to these rules against those who are not. The resolutions of Rio were therefore targeted precisely to this point.

3.2   The Vertical vs. Horizontal Dimension of Sustainability

As mentioned before, the triple bottom line concept is focused on saving the environment for the sake of the present and future generations. This strategy is called here *the vertical dimension* of sustainable development. The addressees of sustainable development are, as a rule, large corporations, which are mostly based in industrialised countries in the northern hemisphere.

Little attention has been paid until now to what is called here the *horizontal dimension* of sustainable development. It refers to different regions in the world with completely different economic conditions and cultural traditions in comparison with the western world. The need for integrating this new dimension has become compelling since all nations in the world have become massively involved in international trade as deliverers of raw material to other countries to be processed and afterwards to be marketed anywhere in the world. Similarly, large business organisations have moved their production sites to where their markets are or where they can benefit from cheap labour, low taxation, and/or less restrictive application of environmental protection regulations. This fairly recent development, termed globalisation, has been pushed forward by new systems of communication, improvements in the world wide transportation systems, abolitions of trade barriers including custom duties, improved systems of quality assurance and the like.

It is well known that the extraction of raw material sometimes has disastrous consequences to the environment and that the import of foreign goods and commodities, for example agricultural products or the introduction of production, which have been developed as a solution to problems of other countries can have negative effects on the domestic economy and as a consequence to the social system of that region. Such measures very often indicate violations of a responsible TBL-policy or a sustainable development. However, they are done in accordance with the government and large parts of the population as a means to escape extreme poverty. The problem discussed here and which signifies the new dimension of sustainable development is how to fight extreme poverty in these countries and give the poor a share of the world's wealth without destroying the regions' culture, their different social structures, and their people's way of life. In his Report to Millennium Assembly, 2000, the UN Secretary General Kofi Annan put this new task in the following words: "The central challenge we face today is to ensure that globalization becomes a positive force for all the world's people, instead of leaving billions of them behind in squalor. Inclusive globalization must be built on the great enabling force of the market, but market forces alone will not achieve it. It requires a broader effort to create a shared future, based upon our common humanity in all its diversity."

3.3   The Relevance of Cultural Diversity

Culture is often understood as the creation of art and its cultivation. In fact this is just one form of its appearance. Culture of a large group of individuals, termed here a nation, is defined here in a much broader sense to include specific ways of communication, thinking, settling cases of conflicts, coming to terms at decision making or on matters of common interest and the like. A nation's language, religion, jurisdiction, governance are therefore elements of its culture. Its origin is largely tradition, transferred from generation to generation by education, i.e. what ever an individual does is based on or influenced by the culture into which he or she was born and to which he or she belongs. Some of a culture's elements are documented as in holy scripts such as the Bible, the Koran, in pieces of art or they are put into codes such as the rules of language or its legislation. Other elements simply exist in the minds of the individuals as for example the way they decide to be dressed.

Culture is constantly subject to changes due to new developments, different interpretation, disregard and eventually deliberate destruction. Driving forces behind such changes originate from inside a cultural group – its results are sometimes identified as subcultures – or from outside. The reception of those changes can vary from unconfined acceptance to strict rejection. No wonder that they can sometimes give rise to tensions even serious conflicts between different groups in a society or between different cultures.

It should be noted that culture as it is defined here, is not a moral category as it was up to the Age of Enlightenment. Since then, tolerance (laissez-faire) as well as romantic fascination with other cultures have come into vogue and are even considered to be one of the highest moral virtues of a civilized person. As long as different cultures were strictly separated cultural interchanges were almost negligible or unnoticed by the majority of people, this tolerance has never been a real challenge.

Apart from certain elements of a nation's culture which are closely related to the conditions of life there is no rationale explanation for a certain cultural appearances, as for example the language spoken there or its religion. From there, the conclusion can be drawn that cultural diversity is a mere caprice of the history of mankind. In fact many people think so; they are neither interested in getting to know or understand other cultures. In short, they believe that cultural diversity is at best confusing and in the worst case the source of serious conflicts. Other people believe that culture is a gift passed on from past to the present and future generations and they are convinced that it can be destroyed but never be restored or substituted with a new and different one. In their view the alternative therefore is cultural diversity or no culture at all. Cultural diversity is therefore considered an asset and the culture of an individual is part of his or her identity and must be protected as part of his or her human rights. From this cause the question is, how cultural diversity and globalization interact or in what way cultural diversity is affected by globalization.

3.4 Globalization and Cultural Diversity

Globalization is a process at the end of which stands as a vision of the so called global village, a perception which initiates for many people a number of positive, sometimes even romantic, associations. A village is often regarded a small settlement where people know each other not only by their names, but also their families, and their educational and professional backgrounds. There is a school, which all young people attend as their parents and even their ancestors did, a church, where people go to on Sundays, shops and a market place where people do their daily purchases. A village gives also the impression of security and calm. Tradition is the backbone of social life.

On the other hand, many people consider a village a place of provincial life meaning narrow mindedness, intolerable social control, a place of no excitement and inspiration. In short a single village is not the place of cultural diversity. Cultural diversity can be found in multitude of villages or in big cities. That is why, many people prefer to life in big cities enjoying the freedom of their own way of life.

The question here is whether the global village in reality is a mega city or a village by its nature. From experience up to now it can be stated that the term of village has been chosen correctly. It is really meant to be a place of one global market, of one global communication system, which works best when using one common language. Unfortunately regional cultural peculiarities cannot be respected; they can only be accepted, as for example a tourist attraction in the museums. Is this scenario of a global village really, what we want? It must be admitted it has some appeal, but the price for its realization may be very high. It cannot be the place of cultural diversity. People might feel that they are no more self determined, but submitted to an unknown culture, which may eventually result in aggressiveness.

Efforts must be made to make sure that cultures can develop smoothly without any distraction. In the following, three items are discussed as they are playing a central role: They are the media, tourism and the agricultural sector.

3.5 Agricultural Sector

As the word already suggests agriculture is part of a society's culture. The diversity of agricultural methods used all over the world could hardly be larger. Whereas in the western states this sector has turned an industry comparable with any other production sector, agriculture is done in other parts of the world the same way as it was in the early history of mankind. Whereas a modern farmer feeds more than a hundred people, the output of a traditional farmer is often no more than is needed for his own family. Accordingly the rural population varies from about four two fifty percent of a nation's total population. The extreme variance of productivity is one of the reasons that farm produce is at the center of global trade. There is no doubt that agricultural product deliveries are in many cases beneficial to large parts of the global population; in many cases even preventing starvation. However, agricultural trade may also have disastrous effects on the agricultural sector in the country receiving delivery, as it can hardly compete with imported produce. In the worst case local produce will gradually disappear from the market. This process will be accelerated when the imports are effected at dumping prices or at no cost at all, due to heavy subsidies from the governments of the exporting countries. The loss of the economical basis of the rural population has of course also negative effects on the social system. Families are torn apart as young people seek work in big cities, mostly without success, and elderly people stay at their homes with no support from their children.

On the other side, intensive farming in the exporting countries can also have serious negative effects. Large scale farming needs large areas under cultivation for best utilisation of the agricultural machines. The resulting mono culture reduces biological diversity dramatically, making the application of pesticides a requirement. The application of fertilisers as well as of weed toxins, heavy use of ground water for artificial watering resulting in lowering of the water tables are requirements for maximum short term productivity but in the medium and long term they are disastrous to the environment.

The conclusion to this development then is, that the agricultural sector should have generally been exempted from the general rules of the World Trade Organisation (WTO). This claim is of course illusionary, but to the extent that it is still possible, sustainable development in countries of low agricultural productivity should be oriented on the principle of help for self-helping. Building on existing agricultural structures policy should be targeted at increasing the productivity for more competitiveness in their own countries. The consequence in the exporting countries should be to take all the externalities into account and to phase out all subsidies. Both measures are needed for fairer trade conditions in the world market. However, it should be kept in mind that agricultural production conditions in certain regions are so unfavourable that competitiveness never can be reached. In those cases it should be up to the national governments to decide whether or not to support the agricultural sector for the sake of landscape conservation.

3.6   Media Industry

The media industry is another economic sector, which is highly sensitive in relation to a nation's culture. Its products are newspapers, magazines, film, television, radio broadcasting, compact disks, online communication systems etc. They are the most important platform for the transportation and interchange of information, ideas, entertainment and – as a mixture of both – of infotainment. They are also used as teaching material for education. The media can therefore be considered a link between a nation's culture as an idea and its population.

Although it is very often denied by publishers, editors and advertising agencies, which is normally done for down grading their own accountability, the impact of the media on the way of thinking of the masses of the population and its behaviour is tremendous. In a free society, the media are therefore termed the factual fourth power after legislative, executive and judicial constitutional powers. Accordingly in illiberal societies, governments keep tight control of the media industry. Fortunately this becomes more and more difficult as radio broadcast, TV-emissions from abroad can be received via satellite almost everywhere in the world. Moreover, the internet has become a world wide medium for communication. The power of the media has been demonstrated in the second half of the eighties of the last century as the collapse of the communist system in Eastern Europe is largely due to the influence from broadcast emissions from abroad. The market of information of a nation should therefore be open to any media product no matter where it originates

At the same time, as the media markets became more and more liberalized, media firms entered foreign countries to start a new business by building up new production capacities or by taking over existing publishing houses. From an economical point of view, this is not unusual, except for the fact that media are a special product. They are generally considered an essential part of a nation's culture and thereby its identity. Such investments therefore should be undertaken wherever it is possible as a joint venture with a local partner. In addition, as part of the media firm's corporate governance, the work of the editorial staff should be based on a special editorial statute, which makes it largely independent of the foreign investor's will.

Media also includes advertisements and commercials. Generally there is not much difference in comparison to editorial information with regard to the access to this kind of information. Unfortunately there is a strong tendency towards standardization and unification of the promotional contents, which relate to the texts as well as to the images. The advertising industry has achieved a high degree of globalization. Unfortunately, the international advertising industry is promoting the western style of living in highly sophisticated way, which brings its addressees in a persistent conflict between the western and their traditional cultures. The advertising industry and its principles should therefore take the enormous effect of their messages into account and decide – again as part of their corporate governance – to bring more respect to the regional cultural specialities.

3.7 Tourism

Tourism has become a mass phenomenon in the last century as a consequence of the process of the industrialization and as a balance to its aberrations. The number of international tourist arrivals is estimated to about 700 million per year. The world wide receipts from international tourism amounted to 462 billion US$ in 2001 or 670 US$ as an average per arrival (http://www.world-tourism.org). The motivations for international tourism can be seen in recreation, cultural, mostly historic interest, experience of different life styles in comparison to everyday life at home, adventure, excitement etc. or – of course – any combination of them.

As long as tourism in a certain region does not exceed a defined upper limit it can be beneficial to the environment and the cultural heritage as it is very often the economic basis for their preservation. It can also be an excellent opportunity for cultural interchange. Unfortunately this upper limit is far below the level the tourism industry needs for attaining at least short term business success. The market for tourism therefore has become extremely competitive with the consequence that every effort is undertaken to be attractive for tourists, which ends in a complete loss of authenticity of the culture of that region. This unfavourable development can be seen at small fishermen's villages at the seaside, which have been turned into bewildering and noisy agglomerations of accommodation and amusement plants. And it can be seen at former mountain villages in the European Alps which have turned into Disneyland like pleasure parks with the mountains as huge playgrounds.

There is no doubt that tourism can be a massive burden to the environment and the culture of a region to where the streams of tourism go and there is also no doubt that tourism as a special type of mobility is part of an individual's freedom. But it cannot be accepted that cultural assets are sacrificed to tourism. It would be disastrous to tourism in particular and mankind in general.

It is difficult to see an end of this development. However, there is some reason for the assumption that mass tourism has achieved a maximum point. First the tourism industry has recognised that its earlier concept based on mass tourism has failed. It is now about to change its strategy towards quality service and to what is called soft tourism. Second, there are many new countries entering the market of tourism, which will be as a consequence no longer concentrated to a few centers, but distributed to many places. In addition, it is necessary to make the ordinary places where potential tourists are living become more attractive during their leisure time. These cities need recreation areas, cultural attractions, places for sports activities, of communication and the like. Such a policy will result in an identification of its inhabitants with their cities and reduce the drive to leave it on every possible occasion.

3.8   Conclusion

Sustainable development is usually understood as all actions to be taken for the increase of economic and social well being without depleting the environment. In this paper the preservation of the cultural diversity has been included as a new dimension. It has been shown that the cultural diversity of mankind is in real danger. This statement applies in particular to the countries of the so called Third World and the emerging countries, which find themselves in a deep conflict between the western way of life and their own traditional culture due to the massive progression of globalization. It has been shown that in policy making cultural diversity is largely underestimated in its relevance as it has not been identified an asset to build on prosperity and peace of the present and future generations. Three fields of economic activity have been depicted as examples of destructive interference into regional specialities of culture. Proposals of how to proceed to meet the requirements of sustainability in these fields have been provided.

References

[1]   K. A. Annan, We the peoples – The Role of the United Nations in the 21th century, New Century – New Challenges, can be found under http://www.un.org/millennium/sg/report/ch0.pdf, 2002.

[2]   D. O. Hall, Food Security: What have sciences to offer? – A Study for ICSU (International Council of Scientific Unions), can be found under http://www.icsu.org/Library/ProcRep/FoodSci/fs.html.

[3]   International Institute for Sustainable Development (IISD): Word Watch Glossary, can be found under http://www.iisd.org/didigest/glossary.htm.

[4]   Saunders, The Problems and Politics of Publishing in the Third World – With particular Reference to Textbook Publishing in Africa, can be found under http://apm.brookes.ac.uk/publishing/culture/1996/SAUNDERS.HTM#LINK1, 1996.

[5]   The SustainAbility People Network: Emerging Economies, can be found under http://www.sustainability.com/programs/emerging-economies/default.asp.

[6]   The SustainAbility People Network: Good News & Bad: The Media, Corporate Social Responsibility and Sustainable Development, can be found under http://www.sustainability.com/publications/engaging/good-news-and-bad-more3.asp.

[7]   World Tourism Organisation: World Tourism in 2002 – Better than expected, can be found under http://www.world-tourism.org/newsroom/Releases/2003/jan/numbers2002.htm

4 Technological Progress in Different Cultures and Periods: Historical Evolution Projected into the Future

Raoul Weiler

Prins Boudewijnlaan, 113, 2610 Antwerp, Belgium, raoul.weiler@skynet.be

4.1 Introduction

Technological progress is considered today as an almost natural phenomenon. In the industrial world everyone lives with the concept, our education system from the lowest to the highest grade is built on it, technological progress is taught as being the guidance for our society. Universities are evaluated according to their scientific and technological performance, their participation in innovation processes and the number of spin-offs they produce. The industrial and Western society has made technological progress as its highest value, an objective to be pursued at any price.

In fact, this evidence needs to be questioned and investigated especially in view of the construction of a sustainable world. It is important to analyze and understand, to which degree technological progress has its roots in our Western culture and not in other major ones. This consideration leads to the major question why the industrial revolution has not taken place in the Greek, Roman, Islamic or Asian cultures, but precisely in Northern and Western Europe. For this endeavor we will follow the extensive and excellent work of J. H. J. Van der Pot of whom the original publication is available in German: "Die Bewertung des technischen Fortschritts. I+II" (The Evaluation of Technological Progress, Part I and II) [1].

In regard with the sustainability axiom, it will be discussed in frame of two centuries the past and present one. Looking ahead on future developments in science and technology a search for a place of technology as a facilitator for attaining a sustainable world society will be discussed.

Global Sustainability. Edited by P. A. Wilderer, E. D. Schroeder, H. Kopp
Copyright © 2005 WILEY-VCH Verlag GmbH & Co. KGaA, Weinheim
ISBN: 3-527-31236-6

4.2   Historical Perspective

4.2.1   Why and Where Technological Progress?

On one hand the absence of a dynamic ambition for technological progress in the earlier European cultures – Greek and Roman – can be explained by their own world vision (Weltanschauung). On the other, which elements can be discovered in the European thinking of the Middle Ages and in the period from the 16th till the 18th century, allowing the future emergence of the industrial revolution of the 19th century and beyond? Technological progress and the birth of the industrial revolution in the Western European countries has lead historians and scientists to look primarily for an interpretation and understanding for the absence of technological progress in non-European cultures. We will look, on the contrary, for the reasons why technological progress has its roots in Western European culture. *It is not the high mortality of some societies or regions of the world does raise questions, but on the contrary, the decreasing and low mortality asks for an interpretation, which appears to be related to scientific and technological knowledge.* It is not social insecurity within given societies that requires an answer, but the social security warranting a descent existence of whole populations as it has been built and acquired in European societies through the benefits of technology, socio-culture and legislative processes, that merits attention and explanation. Many other examples can be added to this list. In conclusion one could formulate it as follows: along the history of mankind, technological progress is rather a concept of scarcity and rarity. This historical fact and situation needs further interpretation and understanding, allowing a better insight and evaluation of the chances we have to reach a sustainable world society in the next fifty years or so.

4.2.2   Religious and Secular Approaches

Among theologians it is today generally accepted that profanation (Desakralisierung) of Nature is as a necessary step which has made the emergence of the modern dynamic striving for technological progress possible. The Anglican Archbishop William Temple (1928) has expressed the difference between Judeo-Christian and other World religions in relation with their attitude towards Nature in the following words: "Christianity is the most materialistic of all higher religions, for while they attain to spirituality from turning away from matter, it expresses its spirituality by dominating matter". In the pagan beliefs, Nature is a motherly deity and the religious beliefs are expressed by the use of magic and sacrifices, allowing the bending of the capricious forces of the gods to their favor. Natural catastrophes are the expression of the rage of the gods. As a consequence technical interventions or constructions, avoiding such catastrophes or reducing their effects, have no chance in pagan society to emerge and would be an attempt of the authority of their gods. The intervention in Nature such as plow the soil, seed and harvest have to be done with extreme caution and accompanied by religious ceremonies.

From the point of view of philosophers, sociologists and other scientists – who not necessarily adhere to a Christian world vision – the significance of Judeo-Christian monotheism in the emergence of the modern technology is as much accepted as by theologians. According to Max Scheler: "The Judeo-Christian monotheistic Creator (christlich-
jüdisch Schöpfermonoteismus) and its triumph on the religion and meta-physics of the
antic world has without any doubt freed the way to the fundamental opportunity for a
systematic exploration and research of Nature. It is an opening of Nature to research and
exploration in a degree which magnitude surpasses by far what has happened up to then
in Western culture (abendländische Kultur). The victory over the natural religions signifies the 'de-personalization' of Nature, consequently Nature becomes an object at the
disposition of human intervention".

And finally, according to R. Hooykaas: "In total contradiction to pagan religion,
Nature is not a deity to be feared and worshipped, but a work of God to be admired,
studied and managed". "The biblical conception of Nature liberated man from the naturalistic bonds of Greek religiosity and philosophy and gave a religious sanction to the
development of technology, that is, to the dominion of Nature by human art".

When the opinions of many researches agree on the positive importance of the contribution of the Judeo-Christian world vision through its 'de-deification' and profanation
of Nature (Desakralisierung), then also many of them remain very critical to the impact
of modern technology on society.

4.2.3 Perception of Nature as an Obstacle to Technical Progress

In the Asian Cultures
In non-Judeo-Christian cultures, the fearfulness towards Nature has been attributed to be
the major obstacle to the breakthrough of a technological grip of man on Nature. Especially this seems to be case for a country like India, with very high cultural development,
in which the cultural vision is founded on the assumption that all living beings belong to
one whole world or system. In other terms, the necessary condition for technological
progress to break through, is that the value of the human being is superior compared to
the other living species in Nature. This perception stimulates the conscious, methodological and systematic domination of Nature by man allowing him the construction of a
technology driven society.

Historically it may appear surprising that, according to Joseph Needham, a country as
China, which had reached around the 14th century a higher technical development than
Europe, and according to Fernand Braudel, China disposed since the 13th century over
the material means to bring about an industrial revolution, did not succeeded to do so.
The fundamental reason has not to be sought in the fact that the Chinese did not have a
sufficient degree of rational thinking, but rather that their rationalism was a different
one. Max Weber wrote: "The Confucian rationalism signifies rational adaptation to Nature and the puritan rationalism a rational domination of Nature".

In the Islam World
In the Islam world, according to Seyyed Hossein Nasr (1976), "In fact it might be said that the main reason why modern science never arose in China or Islam is precisely because of the presence of a metaphysical doctrine and a traditional religious structure, which refused to make a profane thing of Nature…". The most basic reason is that neither in Islam, nor in India nor in the Far East was the substance and stuff of Nature so depleted of a sacramental and spiritual character, nor was the intellectual dimension of these traditions so enfeebled, as to enable a purely secular science of Nature and a secular philosophy to develop outside the matrix of the traditional intellectual orthodoxy. Islam, which resembles Christianity in so many ways, is a perfect example of this truth, and the fact that modern science did not develop in its bosom, is not the sign of decadence as some have claimed, but of the refusal of Islam to consider any form of knowledge as purely secular and divorced from what it considers as the ultimate goal of human existence."

In the Greek and Roman cultures
It has always surprised that in the Greek culture, technology did not receive the place it merited. Indeed, scientific knowledge and systematic and methodological research was very present in Greek society, but these understandings were never brought to practical use in form of technical devices or artifacts. Although the Greeks were disposed of the necessary intellectual tools to come to a technologically oriented society, apparently they stopped at their pre-mechanical understanding and did not advance beyond it. W. J. Verdenius wrote: "According to the Greeks, the destiny of man is to live in the world and not exploit or correct it. For this reason they refused – technology as an activity which subordinated Nature to man and they accepted automates as devices for entertainment and pleasure. The Greeks did not want to mutilate the artistic value of the world by transforming it into an artifact. To the respect of the esthetical value, a religious respect has to be added. Nature is a divine being whose autonomy should not be altered."

The Roman society, although very well aware of its historical mission, was convinced of an eternal cyclic return of the universe and of life on earth. This attitude produced an ideology of resignation paralyzing any perspective for change through technical progress or improvement.

4.2.4   Enlightenment: A Step towards Modern Times

This short introduction provides a glance over the cultural conditions for the emergence of technical progress and the birth of an industrial society in Northern and Western Europe. According to this analysis, the Judeo-Christian world vision contains the prerequisites for the emergence of technological progress and later the industrial society.

The exact sciences contributed substantially to the thinking of the Enlightenment period. The scientific evidence that the earth was not the center of the universe, and consequently that humans created by God and living on it, were not at the center of creation, changed profoundly the perception of the place of the earth within the universe, but more crucial changed their own perception of themselves in the universe and creation. Highly important in the context of these considerations is that the application of a scientific methodology accompanied with adequate observations lead to establishing physical laws, explaining the functioning of the universe and of Nature later on. Scientific rationalism was born and with it the Enlightenment optimism. Knowledge acquired through sciences and the application of their methods, would make it possible to solve many of the problems humans encounter in life on this earth. The novel of Francis Bacon, The New Atlantis [2], paints the most optimistic vision about the power of science and technology

4.3   Modernity and the Sustainability Concept

In view of the perspective of today's sustainable development axiom, the question rises if Western society will have the ability to modify its vision about the meaning and significance of Nature. The predominant anthropocentric world vision of Judeo-Christian religion has to be put into question. Should an eco-centric dimension be reintroduced in the concept of the industrial society, allowing a real chance to access to a sustainable world for the future? Humans are part of Nature and not the master of it, and as a consequence, the technological and economic activities should be subordinated to the ecological planetary conditions. Such a cultural transformation represents de facto a paradigm shift in the Western way of thinking and in Western culture. Does it mean that a necessary but profound step has to be taken in the next half-century or so?

4.3.1   The Twentieth Century

Scientific progress has been one of the most spectacular phenomena of the past century, compared to the preceding ones. The improved understanding of the structure of matter as well as new insights and comprehension of the origin of the universe are remarkable achievements of physics. The breakthrough in the understanding of biological reactions and processes, of the structure of the biological cell and the discovery of the composition and structure of the double helix of the gene have opened new insights about the origin of life as well as about potential engineering of the living world.

The technological progress, regrettably to admit, has been accelerated during and after the two world wars. The living standards of the populations of the industrial societies have reached unknown heights in terms of material comfort, health care resulting in increased aging of the population, transportation and mobility. In the last decade the emergence of the World Wide Web has brought new opportunities for connecting people around the planet at any time and creating access to information and knowledge.

However, some shade sides of technological development and the progress that was accomplished became evident from the nineteen sixties on. Unexpected questions arose in society about their impact on the eco-system of which humankind is part of. Eminent scientists and philosophers, as Rachel Carson [3] and Arne Naess [4], discovered and denounced the negative effects of chemical products and practices of manufacturing companies on the environment. The green movement came into existence. Through a persistent discourse on pollution effects and destruction of the environment, the green movement contributed on large scale to a rising awareness among the larger public. The reports to the Club of Rome [5, 6] about the limits of available resources on earth cautioned for appropriate use. Research was commissioned by governments and international institutions about the carrying capacity of the planet. Scientists were asked for interpretation and understanding of the trends, and political leaders were challenged to decide on appropriate measures to be taken. International and world conferences were called to debate on the environment and the planetary ecosystem. Environmental thinking was the first step in the process of growing awareness of the state of the planet.

Similarly, it became clear that the fabulous technological progress made in the twentieth century had not solved some of the major problems of humankind. Poverty had increased and about three quarters of mankind remain poor, enough food is a daily problem for millions of people and the situation of illiteracy is about the same, etc. In fact the benefits of the technological progress are only for the industrialized countries, which steadily increase their material well being. The overall environment does not improve or gets even worse due to demands of the industrialized societies of resources for their consumption.

The UN General Assembly called in the late eighties for a special commission to write a report on the future of the planet and mankind. That report was titled, Our Common Future [7], in which the new concept of Sustainable Development was defined and elaborated. It was stated that the technology was available to allow the future development to be sustainable, meaning that the future generations would have the similar opportunities to develop as the actual one of the industrialized world. A new paradigm or axiom was announced, a new vision on society was described that should be concretized in the next two or three generations. A sustainable world society must be the outcome.

In addition to this new perspective of sustainability the additional parameter of a quasi doubling of the population by the end of the following century has to be taken into consideration. This is an enormous additional challenge for reaching a sustainable world society in which material goods like housing and infrastructure, goods for well-being like education and health care, goods for immediate survival as fresh water and food are provided in sufficient quantities and on an equitable basis.

Technological progress is imprisoned in a triangle of contradictory tendencies and expectations [8, 9, 10]. Firstly, the eco-system and resource availability require a radical repositioning of the place of technology in all societies: industrialized, emerging and developing ones [11]. Secondly, the inequitable share of technological benefits of the past and present has created high aspirations in emerging as well as in developing societies, especially among the younger generation. Thirdly, the sustainability concept has to include the long term requirements as a consequence of the dramatic increase of the world population, the eradication of present and future poverty and securing a decent life for all humans on earth.

When the picture of sustainability and population increase is not particular attractive, then still, the optimistic vision inherited form the Enlightenment period can be maintained at condition that a radical change of the use of technology is accepted and implemented by society. The UN Conference on Environment and Development, in Rio de Janeiro in 1992 and the Kyoto Protocol to the United Nations Framework Convention on Climate Change of 1997, permitted a real optimism about an agreed change of course of the industrialized nations, however, the aftermath of these conferences does not allow optimism to be sustained.

4.3.2   The Twenty-first Century

Science and technology will evolve during this century in a manner which is not really predictable. However, it remains necessary to reflect on the future, for it is the result of the endeavors of all humans and societies on earth [12]. From the scientific understanding and based on estimations some of today's tendencies can be extrapolated with an acceptable degree of certitude. Some are briefly mentioned:

- The already indicated increase of the world population and the overall aging of it.
- Scarcity of fresh water will be problematic in regions where it was not previously the case.
- Effects of Climate Change in certain geographic regions will become identifiable [13].

Fossil energy resources will decline with the imperative implementation of new and especially renewable energy resources.

Less resource use for manufacturing purposes will be necessary.

Biotechnology and genetic engineering will allow modifications in the biosphere.

It is expected that nanotechnology opens perspectives of new materials.

The existing communication and information technologies (ICTs) [14] will further penetrate in all activities of man: chips will be omnipresent in almost all artifacts, in administration, in health care, in education, in leisure, etc.

Computer power will continue to increase substantially and is expected reach and surpass the 'power' of the human brain [15, 16, 17].

Robots will be present in all human activity and replacing labor of humans.

The World Wide Web will include increasingly artifacts of all kinds, eventually surpassing the human presence.

It is anticipated that the fast evolution of the information and the knowledge society will bring strong modifications in the social behavior within societies as well as that economic activities will undergo severe changes. Especially in the domain of education, the ICTs tools are expected to bring new opportunities enabling a 'leapfrogging' progress in the fight against illiteracy and, thus in the eradication of poverty.

Again, from this short overview of what may happen in the next decades, science and technology will be very much in the center in shaping the future. Unless a sudden collapse would occur, which is, from historical perspective, rather unlikely to happen?

4.4   In Search of a Coherent Evolution of Technology: Past and Future

In the synoptic table (Table 1) an overview is given of some basics attitudes in different periods of time and extrapolated to the end of the present century. The table is inspired from Carl Mitcham's book Thinking Through Technology (page 298) [18], slightly modified and enhanced with more periods, the Middle Ages, the past and present century and beyond.

By introducing the Middle Ages it was possible to refer to the Judeo-Christian world vision and its impact on fostering the concept of technological progress. With the introduction of the recent periods it was possible to incorporate the sustainability concept in the historical process of technological progress.

The four basic attitudes towards technology, chosen by C. Mitcham, have been kept and allow a tentative description for the added periods (Table 4.1).

Volition (transcendence):
The embracement of technology involves the turning away from any divinity and the dissemination of technological progress is made possible through the Judeo-Christian world vision in Western society. The profanation of nature is essential in this process. From the Enlightenment till the end of the present century, representing about a period of five centuries, science and technology were first destroyers of the eco-system but are called to become in the future contributors and later on maintainers of a sustainable world society.

Activity (ethics):
At the personal level: technology and sciences have allowed individuals a great physical mobility as well as an overall interconnectivity, independent from time and space. Ultimately the person will be replaced by robots in the exercise of his/her labor activities.
At the societal level: technology has weakened social cohesion however, is expected to become an enabler for local development and education. A global ethics for equity and solidarity are required for achieving a sustained society.

Knowledge (epistemology):
The breakthrough of the Western scientific methodology opened the door to a new way of thinking: the scientific rationalism. The recognition of the power of human's rational thinking and the discovery of laws of nature enabled the emergence of the industrialized society. Sustainable issues require a holistic methodology able to cope with complex situations. The understanding of the laws of chaos and complexity completes the picture of knowledge about nature and society.

Objects (metaphysics):
Artifacts have acquired daily presence in industrialized societies. However, their manufacturing and availability are a continuous threat to the quality the eco-system. Environmental protection has emerged from their use as well as sustainable development from their demand of resources and raw materials. The concern of climate change is a new phenomenon which will lead to an overall tendency for restoring the eco-system.

Table 4.1 Basic attitudes towards technology in the past and the future (partly after C. Micham (1994). Thinking through technology [18])

	Volition (transcendence)	**Activity** (ethics)	**Knowledge** (epistemology)	**Objects** (metaphysics)
< 0 **Ancient Ages** Skepticism (suspicious of technology)	Will to technology involves tendency to turn away from God or the gods.	Personal: Technical affluence undermines individual virtue. Societal: Technical change weakens political stability	Technical information is not true wisdom.	Artifacts are less real than natural objects and thus require external guidance.
0–1600 **Middle Ages** Innovation (introduction of Techniques)	Will to technology emerges as a result of the profanation of Nature (Desakralisierung).	Personal: Technical craftsmanship as a source arts and wealth. Societal: Technics and tools for local development.	Technical and scientific knowledge diminishes the acceptance of revealed knowledge.	Artifacts & tools replace human effort.
1600–1800 **Enlightenment** Optimism (promotion of technology)	Will to technology & science become an activity of human thinking. Birth of scientific rationalism.	Personal: Technical activity socializes individuals. Societal: Technology creates public wealth.	Technical engagement with the world yields true knowledge (pragmatism).	Nature and artifice operate by the same mechanical principles.
1800–1900 **Romantic Period** Uneasiness (ambivalent of technology)	Will to technology is an aspect of creativity, which tends to crowd out other aspects.	Personal: Technology engenders freedom but alienates from affective strength to exercise it. Societal: Technology weakens social bonds of affection. Exploitation of the labor class.	Imagination & vision are more crucial than technical knowledge.	Artifacts expand the process of life and reveal the sublime.

UNIVERSITY OF HERTFORDSHIRE LRC

	Volition (transcendence)	Activity (ethics)	Knowledge (epistemology)	Objects (metaphysics)
1900–2000 Modern Times (technology for growth & wealth)	Will to technology drives economic development. Economic growth as a solution for development.	Personal: Technology creates mobility and personal freedom. Societal: Social cohesion retrogrades. Unemployment as new phenomenon.	Technology invades society and enables the emergence of information and knowledge society.	Artifacts and industrial processes deteriorate and exploit the ecosystem.
2000–2100 Transition Period (Ubiquitous technology: bio- & genetic technology, ICT, etc.)	Will to technology for contributing to sustainability issues.	Personal: Technology makes the individual independent of time and space. Societal: Unemployment remains structural. Technology as an enabler and empower for local development.	Holistic methodology applied to sustainability objectives. Cognitive & network sciences emerge, helping to understand laws of chaos and network in society.	Non-fossil energy sources are implemented. Climate Change a phenomenon and large scale threat. Sustainability as driver to improve the ecosystem.
> 2100 Sustained Society (technology as a companion)	Will to technology for maintaining the planet going. World population reaches steady numbers.	Personal: Robots replace human labor. Societal: Revival of local communities. Solidarity and global ethics.	Remaining 'indigenous' knowledge is part of the human knowledge base. Cultural diversity & traditions reactivated.	Artifacts are 'intelligent' and embedded in global networks. Rehabilitation of ecosystem.

4.5   Conclusions

Technological progress has its roots in the Judeo-Christian world vision in which one of the fundamental elements is the profanation (Desakralisierung) of Nature. Consequently, an anthropocentric world view emerged, the eco-centric values of the world are abandoned, whereas in other cultures they remained present, albeit in different degrees. In the Middle-Ages the concept of technological progress is stepwise accepted and prepares the advent of the period of Enlightenment.

In the Enlightenment Period the breakthrough of the scientific method of experimental sciences takes place and gives birth of the 'scientific rationalism'. The most optimistic perspectives of the development of sciences and technology for the benefit of humankind are announced. These perspectives will be the fundament for the industrial revolution in Northern and Western European countries.

In the past century the continuous expansion of industrial society has been questioned in relation with the pollution and the destruction of the environment. The concept of the sustainability emerges in order to curb the exaggerated use of natural resources as well as for a more equal distribution of the benefits of technology.

The fundamental question is, will the present society, in the present century, be able to orient the use of sciences and technology for reaching a sustainable world society and transform the sustainability axiom [19] into a certitude matching the physical reality?

In the envisioned information and knowledge society of the twenty-first century, science and technology are in its center. However, their mission should be transformed from a threat to the planetary eco-system into a facilitator to sustainability.

Given the profound roots of technological progress in the culture and world vision of the Western societies with a strong anthropocentric vision, the acceptance of change for reaching a sustainable world society requires a more equilibrated one allowing the incorporation of eco-centric values. A new vision or world order will be necessary. Without any doubt, such a transformation will require a strong and convinced intellectual and political leadership.

References

[1]   J. H. J. Van der Pot, Die Bewertung des Technischen Fortschritts I+II. Van Gorcum, 1985.

[2]   Fr. Bacon, The New Atlantis. Het Nieuwe Atlantis, Ambo/Baarn, NL(1627, 1988).

[3]   R. Carson, Silent Spring. Houghton Miffin Co. 1962.

[4]   A.Naess, in Deep Ecology for the 21st Century (Ed.: G. Sessions), SHAMBHALA, 1995, pp. 64–84.

[5]   D. H. Meadows, D. L. Meadows, J. Randers, W. W. Beherens III, The limits to growth, Universe Books, 1972.

[6]   M. Mesarovic, and E. Pestel, Stratégie pour Demain, Seuil 1975.

[7]   The World Commission on Environment and Development, Our Common Future, Oxford University Press, 1987.

[8]   R.Weiler, and D. Holemans (red.), Gegrepen door TechniekK VIV/ Pelckmans, 1994.

[9]   R.Weiler, and D. Holemans (red.), Bevrijding of Bedreiging door Wetenschap en Techniek. K VIV/Pelckmans, 1996.

[10]   L. Winner, Autonomous Technology, The MIT Press, 1992.

[11]   E.Von Weizsäcker, A. B. Lovins, L. H. Lovins, Factor Four, Earthscan, 1997.

[12]   J. Bindé (Ed.), Les Clés du XXIE Siècle, UNESCO/Seuil, 2000.

[13]   R.Weiler, and D. Holemans (red.), De Leefbaarheid op Aarde, TI-K VIV/ Garant, 1997.

[14]   UNESCO, World Communication and Information Report, UNESCO Publishing, 1999.

[15]   R. Kurzweil, The Age of Spiritual Maschines, Viking, 1999.

[16]   R. Rhodes (Ed.), Visions of Technology, Touchstone, 1999.

[17]   M. Kaku, Visions, Oxford University Press, 1998.

[18]   C. Mitcham, Thinking through Technology, The University of Chicago Press, 1984.

[19]   W. Sachs, Planet Dialectics, Zed Books, 1999.

5   Views of Sustainability: Elements of a Synthesis

Dimitris Kyriakou

Institute for Prospective Technological Studies, Joint Research Centre,
European Commission c/Inca Garcilaso s/n, Edf. EXPO, 41092, Seville, Spain,
dimitris.kyriakou@cec.eu.int

5.1   Introduction

Sustainability and the debate around it raise tensions and has far-reaching implications, because, almost inevitably, it raises the spectre of the market system undermining its own pillars and prerequisites. At the same time, perhaps equally inevitably, it has given rise to an ever-expanding literature as diverse in scope and focus as the backgrounds and agendas of its authors. This would be expected since environmental issues are among the thorniest cases of market failures, second perhaps only to health care policy issues, and invite the contribution of analysts from a wide array of disciplines, for their understanding.

It should be made clear at the outset that underlying this chapter is an anthropocentric emphasis: safeguarding the ecosystem's resilience, i.e. its ability to withstand and absorb perturbations without drastic shifts in its configuration, is important, but not as a goal per se, beyond its impact on human welfare. Since this chapter takes the view that for policymaking and policy-informing purposes the relevant perspective is that of improving the welfare of human society, biological imperatives need to be subjected to the effect-on-human-welfare test. In other words, environmental protection is not treated as an ultimate compass above and beyond human welfare but is rather examined inasmuch as it affects the latter.

Within this context, this chapter will present the main issues and approaches on this topic and will attempt to anticipate elements of a possible consensus. In other words its contribution to the debate will be a distilled version of different views on sustainable development, leading to a view (definition/criterion), which can act as a focal point for consensus-building.

Global Sustainability. Edited by P. A. Wilderer, E. D. Schroeder, H. Kopp
Copyright © 2005 WILEY-VCH Verlag GmbH & Co. KGaA, Weinheim
ISBN: 3-527-31236-6

Briefly, a *synthesis* may entail a two-tiered, weak-strong sustainability criterion, whereby:

i) for areas of human activity, and those natural stocks, for which renewability/ recyclability/substitutability is achievable with presently available means, their de-pletion would carry a – possibly very large – finite cost

ii) If the above condition is not met, then a rigid constraint is set on the consumption of these 'endangered', non-recoverable natural stocks; the violation of the constraint carries an infinite cost.

The above holds if the evolution of the natural resource stock is known to follow smooth patterns; i.e. it does not obey a non-linear model with catastrophe thresholds accompany-ing cataclysmic, irreversible shifts from one 'state' to another (one basin of attraction to another, in the language of authors steeped in the theory of chaotic behaviour). For the resources who do exhibit such *non-linear behaviour*, the resilience of the system in the present configuration is to be examined. As long as we are far from a critical threshold we can pursue an approach of finite depletion cost as in i) above. If not, or if we need more time and research to figure out how close we may be to a threshold, we should adopt a more demanding criterion, approximating ii) above. The rigidity of the constraint will be strengthened (and the costs associated with depletion will approach infinity) or relaxed on the basis of research results, indicating the existence/imminence of a catas-trophe threshold. A course of action in this view should be undertaken not if its expected benefit is greater than zero, but if it is greater than a 'quasi-option' value, reflecting the cost of making it impossible (foreclosing the option) for future generations to decide on implementation, after scientific progress will have shed more light on the issue.

In all cases the goal is to be able to *preserve the set of options for the coming genera-tions*, and not to irreversibly foreclose some or all of them. Furthermore, achieving the goals and meeting the constraints has to be pursued in a cost-effective way. *Note that this is a weaker condition than strong sustainability, which requires the preservation of individual stocks of capital. The criterion proposed herewith allows the reduction of all stocks, as long as irreversible harm (option foreclosure) is not done.*

Another key feature is that although the cost to future generations for restoring stocks to *their* preferred levels may be very high indeed, it is advisable not to fall into the pater-nalist trap and pretend foreknowledge of their preferences. It is preferable to give them the option to choose their preferred level for each stock, as well as the ability to attain it. *To afford them the ability to attain it* (e.g. by developing the appropriate technology), we must try not to foreclose options, to achieve our sustainability goals cost-effectively, promote science and technology in order to give them the scientific/technical basis to go after their goals and allow them enough income to afford their goals.

This latter income-related condition should not be under-appreciated. It brings 'weak' and 'strong' views together in a round-about way. Weak views call for non-declining income for future generations. They should certainly have that and then some, if they are to be more than nominally free to exercise the option to choose and achieve their sustainability goals. As the 'strong' views often suggest, current income may be 'taxed' as a result of regulations promoting environmental sustainability; on the other hand however, the impact of natural-stock-depletion to the income-producing abilities of future generations could be devastating. Emphasising therefore the importance of having the income to 'pay for' the sustainability goals future generations will choose, brings together the 'weak' emphasis on non-declining income, and the 'strong' emphasis on bequeathing to our descendants the natural stock with which to generate income. The two views may be reluctant bedfellows, but bedfellows they are nevertheless, in this case.

In brief a possible consensus-galvanising synthesis could consist in: a) assigning a finite (*possibly very high*) cost to resource depletion, when substitutability obtains; b) treating resource preservation as a rigid constraint whose violation carries an infinite cost, when faced with irreversibility; c) postponing critical decisions for as long as possible, when severely constrained by information scarcity (e.g. on the proximity of thresholds) – thus preserving the option set until scientific/technical progress reduces uncertainty; d) *promoting scientific and technical progress*, not only as a means to *resolve uncertainty* and *enhance substitutability* but also as an *engine of economic growth that will finance substitution across forms of capital, where technically feasible.*

The dual role of technical progress, both for enhancing substitutability and for sparking growth, can not be emphasised strongly enough. Preserving the natural capital restoration option for future generations may prove irrelevant if the technical and economic ability to afford restoration/recovery is not available. In this light, growth and sustainability will either flourish in unison or stagnate in discord.

The rest of this article is organised as follows: section 2 provides an overview of the arena of competing definitions/criteria and the weak – strong dichotomy. Section 3 briefly reviews the aforementioned dichotomy in the context of appropriate accounting to reflect sustainable development considerations. Sections 4 and 5 explore issues and arguments associated with the weak and strong versions. Section 4 reviews the seminal result in the neoclassical literature on sustainability, namely the Hartwick-Solow rule. Section 5 examines the role of substitutability across forms of capital (man-made vs. natural) and the role of technical progress. Section 6 reviews issues related to irreversibility and catastrophe thresholds. Section 7 provides an overview of basic models, section 8 introduces the issue of the discount rate and section 9 closes with a concluding discussion.

5.2   Definitions

As with most complicated issues defining the problem is often half the solution. Often quoted definitions call for development that meets the needs of the present generation without compromising the ability of future generations to meet their needs. This is vague enough to be pleasantly quotable yet not operational. In general, operational definitions offered have been categorised as calling for 'weak', 'strong' or 'very strong' sustainability.

5.2.1   The Weak Version

The 'weak' version calls for guaranteeing non-declining utility [1] for future generations. Its more operational equivalent – given the difficulty in predicting preference sets for our descendants, as well as the focus on consumption-generated utility in economics – is guaranteeing non-declining consumption for our descendants.

An alternative formulation that will facilitate comparison with the 'stronger' definitions calls for a non-declining value of the aggregate capital stock (i.e. the sum of the values of the individual capital stocks, typically classified as man-made, human and natural). Underlying this reformulation is the Hicks – Lindahl concept of true income: the maximum amount that can be spent on consumption in the current period without reducing real consumption expenditure in future periods. This would imply that a part of today's income has to be saved – i.e. not consumed – to replace the amount of capital that was eroded in producing today's income, thus allowing the same level of total capital, production and consumption tomorrow.

5.2.2   The Strong Version

The 'strong' view targets the preservation of each individual capital stock, and not simply of their sum total. Implicitly what is at issue here is the substitutability between natural capital and man-made or human capital (knowledge, technology, etc.) – although some suggest that different forms of capital should be seen as complements rather than substitutes [2]. The 'very strong' forms deny the possibility of substitution across different kinds of natural stock and hence call for the preservation of each kind of natural capital stock. Finally the 'strongest' forms reject focusing on preserving the *value* of the individual components of natural capital, and aim at the preservation of fixed physical levels of the natural stocks in question.

5.2.3   Alternative Views and Discussion

Perrings [3] has put forth an intermediate criterion whereby i) man-made capital is non-declining; ii) the sum of man-made and natural capital (setting human capital aside for simplicity) is non-declining, and iii) in the limit natural capital is non-declining – it may decline for some finite period but not leading to complete depletion. The latter condition follows from the postulation on the part of Perrings that no man-made capital can be produced without the use of a positive amount of natural capital; if natural capital is depleted no more man-made capital can be produced, and therefore man-made capital will eventually decline, violating condition i).

Vercelli [4] suggests as a guide for sustainable development (henceforth SD) the preservation of a set of options at least as ample as the one enjoyed by the current generation as a way to guarantee that choice is effectively preserved. One can add to this the need for growth, and income that will ensure that choice can be effectively exercised; an alternative is that much more important when the agent can afford it.

In all approaches, all of the services, functions, amenities of natural capital must continue to be provided. The underlying question is the extent to which there can be alternative sources for these functions. The 'weak' version has been criticised for being too optimistic on the prevalence of substitutability, and on the reliability of prices and monetary valuations as indicators [5]. A quick summary of examples of environmental degradation blamed on economic activity includes: "depletion of mineral resources, decimation of commercial fish stocks, acidification of freshwater lakes, thinning of the ozone layer, accumulation of persistent toxic chemicals in flora, fauna, sediments and the ambient environment, loss of top soil, accelerated rates of species extinction, deforestation and desertification, and the build up of greenhouse gases in the atmosphere." [6]

Another way to view the weak-strong dichotomy in its starkest form is that sustainability requires the preservation of either the quantity of a stock (very strong version) or its output (weak version) [7]. A way to resolve this is to include all 'amenities' provided by the resource in the 'output' to be preserved, and to guarantee avoidance of irreversible events (i.e. irreversible 'extinction'), because we can not foresee their value for our descendants, and we should maintain the set of options available to them.

The simplification in moving from a non-declining utility to a non-declining consumption criterion is not without its pitfalls. Depending on consumers' preferences GDP growth (and presumably consumption growth) and utility may diverge. In any case environmental degradation may constrain GDP growth in several ways [8] through i) rising marginal costs of extraction, and thus raising input costs per output unit, ii) absolute depletion of some resources, leading to higher prices for remaining stocks or for substitutes (e.g. tropical timber), iii) falling unit productivity of natural resources (e.g. soil degradation in agriculture increasing fertiliser expenditure), iv) higher pollution cleanup costs, registered as higher input costs to guarantee a minimal quality of life.

The aforementioned cases of GDP-constraining environmental degradation can be tackled with 'win-win' policies, featuring no trade-off between growth and environmental quality, i.e. cases in which economic efficiency and environmental protection imperatives point policymaking in the same direction.

Since in any case cost-effectiveness in achieving 'weak' or 'strong' goals is desirable, substitutability should be explored whenever possible. This includes substituting across types of capital stocks (intersectoral substitution), balancing reduction of stock in one locality with increase in another (interspatial substitution), and balancing stock reduction in one time period with stock-restoring frugality in another (intertemporal substitution).

In summary, in this section weak and strong versions have been presented. The differentiating points can be seen to be the extent of substitutability between natural and man-made capital – with the weaker versions accepting and expecting it and the stronger versions strongly doubting it. Intermediate approaches may allow – explicitly or implicitly – reduction of natural capital but setting limits to such depletion (towards preserving the set of options for future generations, or meeting the natural capital requirements for the production of man-made capital). In win-win cases, where environmental and efficiency concerns coincide, there is no tension. Tension arises when activities promoting economic growth (at least in the short term) are inimical to the environment, and to the longer term outlook for income.

5.3   Accounting for SD

Devising new accounting practices to reflect sustainability considerations is far from simple. It entails natural scientific considerations, regarding the continued supply of environmental services, and economic analysis regarding the demand for such environmental functions. Moreover, intertemporal valuation becomes problematic due to the extremely distant time horizons involved. Furthermore, there is no market for environmental goods in general, to allow us to observe their prices directly. Calculation of shadow prices would be needed but the latter would require knowledge of supply and demand characteristics for the goods in question. It is possible, though not easy, to estimate supply curves from the cost estimates of restoring/preserving an environmental function. Demand curves however are nearly impossible to infer because consumer preferences can be only partially traced [9].

Another way to approach the problem is to use environmental standards that agents have to meet. Which standards however and who is to designate them? As a working hypothesis we can accept that the standards chosen by the UN reflect society's preferences. Accounting[1] then would proceed as follows: a) determine the environmental burden in physical units, b) compare with physical standards, c) estimate supply curve from restoration activities/costs, d) estimate minimal costs (cost-effectiveness) in achieving standards. Note that this makes the strong assumption that there is no cross-jurisdictional pollution or that measures are universally applied.

The distinction between 'weak' and 'strong' SD has a parallel in environmental accounting: 'fundist' vs. 'materialist' approaches. 'Fundism' accepts substitutability and is associated with the 'weak' version, and the reverse holds for 'materialism'. In the 'fundist' view, as in El Serafy [10], comprehensiveness of measurements is deemed unattainable; flow variables are emphasised over stock variables due to the latter's problematic pricing; no value is placed on the stock of natural capital but changes in physical dimension need to be valued at prevailing prices (methods used may include indirect valuation – e.g. of products lost by soil erosion, of the cost of fertiliser necessary to compensate for erosion, etc). The emphasis is on proper measurement of income, e.g income resulting from capital disinvestment is not included in the true income category.

This involves key distinctions: in accounting for economic exploitation of natural resources the difference between sale value and cost of inputs is often wrongly identified as rent. The aforementioned difference is a mixture of proper rent, a return to the 'powers of the soil' (also called Ricardian rent), and also a component reflecting the opportu-

---

[1] There are two approaches to environmental degradation cost estimates. The first approach focuses only on primary costs (of restoration, preservation, etc.). The second approach calculates estimate costs by simulating what would happen if environmental services entered the market. The latter approach is plagued by assumptions of questionable viability. The former is compromised by the possibility that a decrease in primary costs may not necessarily reflect corresponding changes in physical parameters (e.g. primary costs may drop due to technical change).

nity cost of depleting the resource (the latter component was called 'royalty' by Marshall and 'user cost' by Keynes). Ricardian rent is legitimately includable as true and sustainable income whereas Marshallian 'royalty' is not. El Serafy suggests that the 'royalty'/'user-cost' part of revenue be reinvested to guarantee non-declining income. Depletable resources are inventories being used up. Their depletion affects the possibilities for GDP generation, and not simply the net national product through simple depreciation.

For renewables El Serafy adopts Daly's [11] recommendation to extract at a level dictated by nature's regenerative capacity. Beyond that point revenue is not true sustainable income. The same holds true for the waste-assimilative capacity, which can also be seen as a depletable asset. The difference between 'fundist' views and 'materialist' views is highlighted in their proposals on the fate of the user-cost. The former suggests reinvesting it with financial criteria; the latter proposes reinvesting specifically in renewable substitutes of depletable resources.

Implicitly again the underlying difference is whether substitution between man-made and natural capital is achievable. If it is, then reinvesting towards maximal financial return is sound advice; the process will build up man-made capital and if substitutability applies then amenities provided by the depleted natural capital will not be missed. If however such amenities can not be replicated by man-made capital the 'materialist' approach is the safer one.

In this analysis of proper accounting of true income the need for appropriate environmental accounting may be stronger for less developed countries (LDCs). The latter are more likely to have high reliance on depletable natural resource extraction. Hence they need to realise quickly how ephemeral their growth may be, and how they may get caught in a low environmental quality, low income trap.

A few technical remarks may be in order before closing this section. Firstly, and quite obviously, natural capital erosion in the preceding analysis includes renewable, non-renewable as well as non-material goods (e.g. biodiversity). Furthermore, accounting tools such as market valuation, prevention costs calculation, contingent valuation, etc. are meant to be complementary ways to impute monetary value for physical-unit accounts, often called satellite accounts (obviously taking care to avoid double counting).

Environmental accounting indirectly involves moral judgments: for instance applying the 'polluter-pays-principle' in environmental accounting is not simple. It is the consumer demanding the good that is charged with the onus of the polluting activity (and its cost). The reason for this accounting attribution of the cost of pollution to consumers and not producers is to avoid the 'exporting' of environmentally harmful activities to other countries (e.g. LDCs) – from which the products of such activities can be imported by willing consumers (e.g. in developed countries). If the onus were not on the consumers, then consumers' demand in developed countries would still drive environmental pollution, and there would be no attribution of the environmental consequences and their 'accounting cost' to this type of demand.

Finally cost-benefit analysis (CBA), a pillar of much accounting work, has been criticised as problematic in its application to SD [12] Beyond the well-known problems of pricing non-marketable alternatives, the multigenerational time-horizon and the uncertainty regarding preferences, system stability, etc. make the application of CBA particularly difficult.

In closing, the weak-strong dichotomy has its analogue in fundist vs. materialist views of sustainable development accounting. In general the standard national accounts are corrected by taking into account the reduction of depletable natural assets (often in satellite physical-unit accounts, which are usually eventually monetised). There are three crucial issues: i) whether prevailing prices are used to monetise physical-unit accounts; ii) how natural capital depletion (capital disinvestment) is to be treated, i.e. what should be entered as true income; iii) what is to be done with the Marshallian 'royalty' (or Keynesian 'user cost') part of the revenue from resource depletion. Should it be reinvested with financial return maximization criteria, or should it be reinvested specifically in renewable substitutes of depletable resources – substitutability is lurking beneath this dilemma.

5.4   The Hartwick-Solow Rule

The weak version is epitomised by the Hartwick-Solow (HS) rule[13, 14] – the focal point of neoclassical treatments of SD. The HS rule calls for investing the scarcity rents (the 'user-cost' discussed earlier) received from the extraction and exploitation of depletable resources, in order to build up reproducible capital to a level that would balance the natural resource depletion and would guarantee non-declining consumption over time – which would be equivalent to keeping the capital stock non-declining over time.

A few observations need to be made in conjunction with the HS rule, regarding its applicability (cf. Solow [15]). First, the HS rule implicitly assumes substitutability, allowing the accumulation of man-made capital to replace – and provide all the amenities of – the depleted natural capital. Second, observance of the rule becomes infeasible when the scarcity rents received are less than the amount necessary for sustainability-guaranteeing capital formation. If we allow slight departures from the idealised setting, we observe that competition among geographically dispersed agents with property rights for stocks of the resource may drive the scarcity rents down to the levels agreeable to the most myopic among them. The rents in this case may not suffice for the rest of the 'producers' (if this storyline sounds familiar, it is because it mirrors the OPEC disputes on oil production). Even without such departures [16] suggests that too fast depletion may lead to rents too small to achieve SD.

Furthermore the assumption is made that intertemporal discount rates and returns to capital will not diverge, i.e. that arbitrage will be adequate to equalise them. Due to the length of time involved however, the arbitrage market may be too thin to achieve such price convergence. If intertemporal discount rates are higher than the capital rental rates, at which HS rents are reinvested, the HS rule may be insufficient to guarantee the required capital formation. Furthermore, as an overall condition savings must cover not only natural capital depletion (a la HS), but also manufactured capital depreciation [17].

Third, the use of Cobb-Douglas constant-returns-to-scale production functions by HS guarantees that the elasticity of substitution between inputs is equal to one, at the expense of keeping technical change exogenous. It would be worth exploring what happens when technical change is made endogenous and when we assume increasing returns to scale. Fourth, the HS model relies on an unrealistic infinitely-lived representative agent framework, not treating explicitly future disconnected generations [18]

Finally, Cobb-Douglas production functions imply that as the depletable resource tends to zero, its marginal productivity tends to infinity. As the available quantity however goes to zero, it becomes increasingly likely that oligopolists or a monopolist will control the market. If oligopoly prevails price trends will depend on whether oligopolists compete a la Cournot (i.e. on quantities), or a la Bertrand (i.e. on price). In a monopolistic setting factor prices will not equal marginal productivity values. In any case if control of the market falls within a single jurisdiction, thorny questions arise regarding the owners' time horizons, their attitudes vis-à-vis the rest of the world and investment patterns (an illuminating precedent for study may be provided by the case of oil and leading oil-producers).

In conclusion, the HS rule is a succinct summary and policy guideline for the weak version. Small departures from its idealised setting however show that the robustness of the HS result is limited. Such departures may lead to rents that are too small to achieve SD. Insufficient arbitrage may allow divergence between intertemporal discount rates and return on capital. As the available quantity of the harvested resource is reduced the probability of monopolistic behaviour increases and is further complicated by political/jurisdictional considerations. Finally the model's reliance on exogenous technology and infinitely-lived representative agents are useful simplifying analytical steps, but could become more realistic through exploring different ways to account for distinct generations and endogenising technical change in the model.

5.5 Substitutability and the Role of Technical Progress

Substitutability is the key issue separating 'weak' and 'strong' versions of sustainability. The latter view promotes the preservation of natural capital per se, whereas the former focuses on the preservation of aggregate total capital (natural and man-made) in the sense of preserving the ability to generate the same services, income, or level of utility, depending on the formulation. The problem is markedly easier in the case of renewable resources where consumption can be reduced to regeneration-conforming levels, or when renewability can be enhanced through human ingenuity.

For both renewable and non-renewable resources technical progress is a critical variable which should properly be seen as endogenous to the economic system. The endogeneity of technical progress reinforces the 'weak' view. Technical knowledge can be seen as a form of human capital which society develops (invests in) out of its savings. It depends not only on the distribution of income among various consumption/investment activities but also on the level of income itself. If income growth is limited due to stringent SD constraints then reduced amounts will be available for human capital formation, which fosters technical progress, as well as for research and technical progress stimulation. The capacity of technical progress to 'clean up after the human race' can enhance the sufficiency of the 'weak' version by allowing man-made capital build-up to balance the gradual depletion of natural capital; it can support SD by allowing substitution across different forms of natural capital by decreasing the burden for the sink function of the environment. Technical progress underlies both the part of economic growth not attributable to growth in the stocks of production factors, but also our greatest hopes for dealing with environmental degradation.

Some argue that historically technological progress has been associated with accelerating environmental degradation. This argument however treats technical progress as an exogenous, deus-ex-machina variable directing growth patterns, a legacy of XIX century potential energy models not conducive to building a positive theory of production processes. Technical change should be viewed as endogenous – as manifested in the endogenous growth, neo-Austrian and neo-Ricardian literatures[19]; it is guided by funding and hope for profit on the supply side, and by consumer preferences on the demand side. Past technical progress reflected in marketed products/processes could not reflect environmental concerns since the latter were hardly salient in consumers' preferences.

Endogeneity requires the remuneration of technical progress (represented by knowledge capital or by knowledge embodied in capital goods), and the introduction of increasing returns to scale in the growth models. Rigidities in technical systems not only make adjustment slow but also punctuate it with tensions during the transition process [20].

The process of technical change is neither smooth nor a-historical. Certain choices may lead to techno-economic paradigm lock-in [21]. Kemp[22] explains how paradigm shifts may be far from painless and may indeed require full-scale revolutionary 'conversions' such as those often accompanying paradigm shifts in science. In light of the above, and given the multiplicity of barriers (economic, technical, institutional) facing new technologies, more radical technology options carrying high environmental benefits in the long run often face an uphill struggle. They need to exploit market niches, creating constituencies behind new markets and new products.

Technical progress can drive economic growth and alleviate social tensions undermining social sustainability. Moreover technical progress can foster substitutability not only in the direct production input role (of mineral resources, for instance) but also in the sink role of the environment in absorbing waste, as well as its role as a habitat.

On the other hand, there are life-supporting functions (food production, protection from ultra-violet radiation, etc.) which can not be fully and independently performed by man-made capital. Second, ecosystems rely only on sunlight and are otherwise self-organising and self-preserving, whereas man-made capital gradually depreciates, making it less reliable. Third, man-made capital requires natural capital for its construction whereas the converse is not true. It is also argued that substitution even across different types of *natural* capital is limited [23]

In conclusion as long as we can afford to wait for technical progress to improve substitutability, our capacity for substitution is likely to ameliorate. The weak version's optimism is supported by the endogeneity of technical progress, viewed as a form of human capital in which society invests – in a virtuous circle technical progress leads to economic growth allowing higher human capital formation expenditure. The situation becomes problematic when dealing with possibly irreversible depletion/damage to the environment, and with choosing the appropriate policies when such an irreversibil-ity/catastrophe threshold is near. Finding a solution is rendered even more difficult in the presence of hard uncertainty (non-diversifiable, non-reducible to known functional forms and probability distributions).

5.6   Catastrophe Points

A substantively rational representative agent maximising his expected utility, under given constraints, is the standard protagonist of most analyses of sustainability. This setting is a-temporal, enjoying the reversibility and function continuity and smoothness reminiscent of XIX century physics, which informed the development of 'marginalist' economics in the last one hundred years. The path followed is not deemed to irrevocably circumscribe present choices, or actually proscribe some of them. The optimising agent is portrayed as making choices with no reference to what has already transpired, even if some of the 'building blocks' are already extinct. Thanks to substitutability and repro-ducibility the amenities provided by natural resource X can be afforded through appro-priate combinations of Y and Z.

Sustainability requirements may be highlighting the limits of this approach by under-scoring the importance of critical thresholds, catastrophe points, as well as ultimate time, energy and matter limitations. Leaving aside the thermodynamic constraints, relevant in the very long run, one can easily object that such predilection for 'smooth', 'well-behaved' functional representations is unrealistic. Phenomena are often more realisti-cally portrayed by non-linear, discontinuous approximations which also include critical thresholds or catastrophe points.

For local analysis, nevertheless, of the continuous, 'well-behaved' functional seg-ments between critical thresholds, the standard optimising framework can be used. For global analysis however the possible existence of multiple equilibria (some less desirable than others) and discontinuities must be seriously entertained, especially when the resil-ience of the system and the proximity of thresholds are not known – where resilience of a non-linear system indicates the ability to absorb shocks without a drastic change in its expected behaviour.

The ecological problem is to determine the minimum combinations of populations that will allow the ecosystem to function with high resilience – avoiding crossing catas-trophe thresholds. In models such as in Perrings [24] multiple equilibria exist; dynamics and stability vary non-linearly with scale; and change is not smooth but rather punctu-ated by sudden reconfiguration, reminiscent of the punctuated equilibria model of bio-logical evolution.

This however is still an optimistic – albeit non-linear – scenario, optimistic in the sense that we have a, possibly complicated, 'map' of possible paths, indicating location and severity of pitfalls as well as our own starting position on the trail. A more pessimis-tic approach suggests that we know neither our starting point, nor the characteristics of other likely candidate destinations (competing equilibria, or basins of attraction). In simpler terms we know neither where we are, nor where we may be going. Not knowing where we are implies inability to estimate the present resilience of the system. Not knowing the characteristics of the possible destinations (i.e. where we may be going) implies lack of predictability; the topology of new basins of attraction can not be inferred without observations, i.e. we will not know where we are before we get there. New dynamics may have to be seen before they are understood, by which time it may be too late (e.g. the vicissitudes of fish stock in US lakes, the management of dry land systems, etc.)

The critique can go even deeper to the heart of the optimising framework, i.e. substantive rationality. The issue is important enough to deserve more extensive analysis. Substantive rationality posits an optimising agent maximising his expected utility, under given constraints. When faced with uncertainty, the agent is assumed to have subjective probability distributions for possible states of the world. He is also assumed to be facing exchangeable events (a-historically independently distributed) [25]. Note that this condition implies perfect reversibility, since reversibility would assign zero probability to certain events after the threshold has been crossed. The set of options facing the agent is not supposed to evolve in time; in fact most often states-of-the-world are supposed to be imposed on the agent exogenously – a stipulation analogous to the exogenous view of technical change. Obviously however, in the case of sustainability agents' actions have a dramatic influence on the eventual state-of-the-world obtained.

The agent is assumed to have a well-defined objective function and knowledge of the constraints facing him, as well as knowledge of the set of feasible choices and of the consequences of his actions. Given all these assumptions the problem can be reduced to one amenable to standard optimisation techniques. Finally, and most crucially, substantive rationality demands that the agent does not commit systematic mistakes. In other words, his failure to exactly predict ex ante (before the resolution of uncertainty) the value obtained ex post (after the resolution of uncertainty) should be due only to the non-systematic, stochastic component of the prediction. The resolution of uncertainty should not make him change the model he used to come up with his ex ante prediction, in order to chose a course of action the next time he is faced with the same stochastic set-up.

This last condition however is violated when we face hard and not soft uncertainty – a distinction muted by many theorists and originally proposed by Keynes and Knight in 1921. When the parameters, the functional form and probability distributions governing the resolution of uncertainty are known ex ante, we are indeed in the realm of soft uncertainty, where we can form a rational expectation ex ante and wait to see ex post which particular value from the *known* probability distribution will obtain for the stochastic variables. When, on the other hand, we are not certain of the nature of the parameters, functional form and probability distributions, we are in the realm of hard uncertainty. In the latter case ex post observations may give us crucial information about the systematic part of the model (e.g. about the topology of the new basins of attraction as mentioned earlier in reference to dynamic models). The *systematic* part of our prediction will be changed then, since we will know more about the model; our original ex ante prediction will prove to have been *systematically* mistaken, in flagrant violation of substantive rationality.

The standard formulation of utility theory blurs the distinction between objectively known probability distributions – for which the original von Neumann – Morgenstern theory of rational decision was proposed – and the subjective probability distributions, for which Savage's formulation was made (note in any case that the latter is seriously undermined when ex ante choice and resolution of uncertainty are far apart in time – as is the case in sustainability-related choices).

Most analysts bypass this vagueness by suggesting that agents do not have to know the objective probability distribution; it is enough to posit that they use their subjective probability distributions and act *as if* they knew the objective ones. It is perhaps edifying to pursue this a little further. There must be a mechanism guaranteeing that agents will come up with the right answers, that they will indeed act *as if* they knew. According to many economists this mechanism is a Darwinian one: those who fail to act *as if* they knew will fall by the wayside. At a micro-level this implies firms going bankrupt; at the giga-level of SD it would imply environmental catastrophe or extinction – the costs of learning from trial and error may indeed be colossal.

It is precisely due to these problems in applying substantive rationality and optimisation to SD that alternative approaches have been suggested, such as procedural rationality, focusing on the procedure employed to reach a satisfactory, though not necessarily optimal, result. The problems in identifying a unique optimal path has also led authors [26, 27] to suggest preservation of the set of options for future generations – instead of a specific 'weak' or 'strong' numerical recipe. Similarly, Hediger [28] argues that the uncertainty about the local impact exacerbates the situation and calls for decision making that takes irreversibility into account such as the minimax regret principle (whereby actors minimise the maximum regret they may experience as a result of their actions).

As Azar & Holmberg [29] explain Rawls proposed that economic optimisation should follow a certain order, granting life, health and liberty priority over consumption. Though criticality will likely not be a problem for all types of natural capital [30], it is cogently argued that catastrophe-risking activities should be avoided if scientific research can clarify the costs and benefits at the end of a certain time period. A project in this view should be undertaken not if its expected benefit is greater than zero, but if it is greater than a quasi-option value reflecting the cost of foreclosing the option of future generations to decide on implementation, after scientific progress will have shed more light on the issue.

In closing this section, the problem in simple economic terms stems from the inability of price signals to reflect the imminence of threshold-crossing, due to sporadic scientific and popular awareness of what these thresholds are; due to distorting policies; due to the public good character of many environmental amenities; and also due to the structure of property rights, leading agents to ignore the repercussions of their own actions. When there is hard uncertainty about the stability characteristics of the system the application of standard, substantive rationality becomes problematic. In the standard economic explication agents are supposed to act 'as if' they knew the underlying objective probability distributions, or else they 'fall by the wayside'. In the case of SD however, 'falling by the wayside' may mean environmental catastrophe.

5.7   Summary Classification of Economic Models

Models of sustainable development come in many flavours. Early models, employing the assumption of substitutability showed that adequate technical progress was able to guarantee SD; technical progress was deemed exogenous and hence attainment of adequate levels was exogenous as well. Simple models examined utility maximization for an infinitely-lived representative agent; others increased realism by assuming overlapping generations and bequests. In most cases the value to be optimised is the value of future utility discounted to the present (net present value – NPV), an approach which obviously places more emphasis on the well-being of the current generation. A Rawlsian maxi-min criterion based on the maximization of the welfare of the least well-off generation has also been proposed. More symmetric treatments of generations reject discounting-to-the-present opting for discounting back to the birthdate of each generation [31, 32]. The latter approach has the advantage of providing a consistent execution of an optimal policy plan across time and generations [33]. Other approaches avoid having to choose between an overwhelming emphasis on present utility (a discounted present value term) vs. one on future utility (an undiscounted long-run utility-level term) by maximizing weighted sums of both terms [34].

There are drawbacks to discounting back to the present, and not only in terms of penalising individuals to a lower level of welfare simply because they were born later rather than earlier. For instance, in the absence of intertemporal markets for environmental degradation, future generations may be disadvantaged, as shown by Mourmouras [35]. Pezzey argues [36] that internalising externalities may be insufficient to achieve sustainability – which suggests that sustainability and environmental policies are conceptually distinct. Moreover, Howarth and Norgaard [37] point out that private altruism and transfers from parents to children may still not achieve a social optimum.

Key for overlapping and non-overlapping generations models (OLG and NOLG respectively) is the specification of the inter-generational social welfare function (ISWF). Pezzey[38] introduces a higher degree of verisimilitude in the models – and also a 'genetic' rationale for SD – by doing away with the parthenogenesis assumption of most models. This latter assumption ignores the fact that humans mate to reproduce and that each parent can claim only half the genes of each offspring. Furthermore parents have no control over who their descendants' mates will be. In this biology-informed view the reason for caring for future generations depends on the extent to which gene preservation can be achieved. Hence, in this view, the gene-pool dilution effect reduces parents' concern for future generations.

Yet another conclusion of the literature is that capital-poor countries may 'optimally' drive themselves to environmental catastrophe. The existence of multiple sovereignty exacerbates the complexity of the problem by raising distribution, trade and pollution rights apportionment issues [39].

Alternative approaches include the neo-Austrian view, which stresses the vertical time structure of production, and hence the relevance of time and the possibility of irreversibility. In this view capital goods creation or productive structure modification takes time; capital is a productive service which is not permanent through time. Neo-Ricardian theory on the other hand views capital goods as productive funds. By considering capital in terms of its environmental resource contents and using Sraffa-Leontieff type input-output analysis, the flexibility, complementarity and environmental-content aspects of the economy can be highlighted [40, 41].

Non-linear frameworks are also used to study the system-dynamic aspects of SD. They often point out the multiple equilibria, basins-of-attraction associated with the problem, as well as resilience issues and the possibility of crossing catastrophe thresholds. Stability analysis then focuses on finding stability corridors in which SD strategies should keep the economic system [42]. Taking their cue from resilience/irreversibility considerations another group of models stresses the hard uncertainty surrounding substantive rationality. They espouse a precautionary attitude emphasising the preservation of the option set for future generations [43].

In brief, economic models for the study of SD vary in their choice of protagonist (infinitely lived single agent, overlapping or non-overlapping generations); in the optimality criterion (net present value discounted back to today or to the beginning of each generation's lifetime, or the maximization of the welfare of the least well-off generation). Most models do not take into account mating and gene-pool dilution criteria, which may reduce agents' concern for the future generations. Finally quite a few models use non-linear frameworks and focus on system resilience, and the identification of stability corridors and catastrophe thresholds.

5.8   Discount Rate

The issue of the appropriate discount rate (DR) has received much attention largely because over long periods of time – such as the ones relevant in this case – a slight change in the discount rate can have dramatic effects for the advisability of certain actions. Most often the problem of sustainable growth is examined as a problem of optimisation of net present value of a stream of 'incomes' taking into account sustainability considerations (the DR problem exists independent of the 'weakness'-'strength' of the criterion used). Obviously a high DR will place larger weights on the utility of the current generation and progressively smaller weights on future ones.

Although the issue has a strong ethical dimension there are two important and often neglected economic observations that need to be made. First, it has been shown that *both* high DR and low DR may justify faster resource depletion. In the case of high DR [44] this is quite straightforward. But even for low DR [45] analysis suggests that lower DR, and hence lower capital costs, reduce the cost of resource-harvesting capital and hence may accelerate resource harvesting and depletion. It can also be argued that the overall economy-depressing effect of higher DR (an income effect in standard welfare analysis terms) may reduce economic activity and hence resource-harvesting.

Diedrich[46] presents an alternative (substitution-effect) explanation of how lower DR may lead to faster resource depletion by arguing that the DR is both an intertemporal price, as well as an intra-period indicator of the relative cost of capital compared to other factors of production. A decrease in DR may reduce the attractiveness of harvesting more in this period, but it may also make capital more affordable and thus change the structure of production towards more resource-depleting input mixes. If the latter effect predominates (and Diedrich gives a numerical example where it does), then low DR may not be environment-friendly. Though in practice the intertemporal DR and the intra-period cost of capital may differ (e.g. due to insufficient long-term price-equalising arbi-trage) these caveats are in principle relevant.

The second observation to be made regards the DR's composition. The criticisms levied against positive discount rates (that they reflect human impatience at best or ra-pacity at worst, that they are an unethical way to value the present generation more than the future ones) apply only to the pure-time-preference component of the DR. The DR however may be positive even if we set pure time preference to zero, because the DR also reflects the rate of growth of productive capacity, and the declining marginal utility of consumption as income grows.

More specifically consumption foregone at time t becomes capital (through the sav-ings=investment link) leading to increased production at time t+1. That capital has to receive remuneration equal to the value of its marginal physical productivity, which implies positive payment to capital, i.e. positive discount rates. Furthermore, increased production and income at time t+1 imply that the consumption postponed at time t in order to be enjoyed at t+1, will provide a smaller marginal utility at t+1, than the same amount of additional consumption, if it were to be enjoyed at time t. This is precisely because the marginal utility of consumption falls as income grows between t and t+1.

In closing, interest rates – the real world proxy for DR – seem to be an inappropriate focus for SD policies, not only because of their ambivalent impacts on the environment as explained above, but also because real interest rates are increasingly viewed as en-dogenous to the economic system and not exogenously set at 2–3% as believed in the past. Recent theoretical attempts limit the importance of DR by focusing on credibly agreed upon and consistently executed policy plans [47]. Such models treat generations symmetrically, discounting utilities back to the birth date of each generation, and not to the present date, as done in the standard net present value (NPV) approach.

5.9   Discussion

A few concluding observations may be in order. First, the crucial issue in this as well as in all welfare economic problems is the specification of the welfare function, both in terms of intergenerational weights and in its treatment of the SD criteria. If natural capi-tal preservation is included as an argument of the welfare function, then its depletion carries a finite cost – substitutability and technology optimism are at work. If, on the other hand, preservation is treated as a constraint then its violation carries infinite costs, and the 'strong' version is espoused.

Second, the overlapping generations models with symmetric discounting back to the birth date of each generation seem theoretically most promising. They should be ex-tended to include endogenous technical progress (and increasing returns to scale) and the gene-pool dilution considerations springing from the fact that humans reproduce through mating and not through parthenogenesis.

Third, a disturbing characteristic of many 'strong' version approaches is the casting of the human species in the role of the alien, extra-natural element [48]. It would be much more realistic, and productive, to view human beings for what they are; namely, members of the animal kingdom whose intelligence, creativity and emissions are part of nature as much as the machinations – albeit less spectacular – of any other species.

Fourth, thermodynamic limitations apply in an 'enveloping' sense in the very long run. Although they are valid constraints to some economists' cosmic optimism, and they do outline the irreversibility of time, they have ambivalent connotations when applied to smaller scales. Besides, Keynes' aphorism that in the long run we are all dead highlights the irrelevance of such giga-scale constraints to economic agents' calculations.

Fifth, the pure-time-preference component of the discount rate indeed raises thorny ethical questions. Nevertheless, given the existence of other components in the discount rate, the endogeneity of real interest rates, and most crucially, the ambivalent effects of discount rate changes on SD, manipulation of interest rates (the real life proxies for the discount rate) is not a useful tool for SD.

Sixth, environmental problems may be exacerbated as less developed countries grow, at least until they reach the turning point on their Kuznets-type curves [49] – i.e. until their income has reached a point where they can afford to take pollution into serious consideration and act on it. Note however that recent research [50] indicates that the samples chosen to examine such Kuznets-curves may play a crucial role regarding where the turning point lies (when developing countries are sampled, as opposed to OECD countries, the turning point in terms of per capita income is extremely high). The political dimension of the problem comes again to the fore in apportioning pollution permits for the commons, in controlling transboundary damage, in distinguishing protectionism from valid concerns for trade-enhanced environmental degradation. If we add to this the possibility of environmental extortion from the 'have-nots' it becomes clear that the environmental and social aspects of SD can not be easily separated.

Seventh, the 'weak' and 'strong' versions offer valuable precepts, which, this paper argues, are ripe for a synthetic/eclectic reinterpretation. The Hartwick – Solow rule, the focal point of the 'weak' approach should be accompanied by several caveats, explicitly mentioned in this chapter (the possible infeasibility of its application, and its assumption of substitutability being the two most important ones). Nevertheless it provides a useful starting point and a facilitating rule-of-thumb for those cases not plagued by the prospect of irreversible damage.

In a sense the 'weak' version is a necessary condition for the attainment of SD; it can approach sufficient condition status as technical progress and scientific breakthroughs expand substitutability (e.g. bioengineering applications), thus repelling the spectre of irreversible damage, and achieve a clear delineation of the trail we are on, the gravity and proximity of alternative outcomes, and the degree of system resilience (i.e. as science transforms hard into soft uncertainty).

In the meantime it may be advisable to combine application of the HS rule with observance of critical thresholds where applicable. Since our knowledge regarding such thresholds is limited, it may be advisable to postpone foreclosing options for as long as possible, in order to give time to science to improve our understanding of what is at stake in each case. In other words, for non-renewable threatened resources, it makes sense to pay the cost of this postponement which corresponds to 'purchasing' a quasi-option allowing us to exercise it, or not, when the scientific evidence is reassuring. The cost of this quasi-option is the immediate benefit we forego in order to preserve the set of alternatives for the future generations.

A course of action in this view should be undertaken not if its expected benefit is greater than zero, but if it is greater than a 'quasi-option' value, reflecting the cost of making it impossible (foreclosing the option) for future generations to decide on implementation, after scientific progress will have shed more light on the issue.

The emphasis on preservation of the set of alternatives – intermediate between 'weak' and strong' versions – has two advantages: first, unlike most 'strong' approaches it is well rooted in recent mainstream neoclassical literature (i.e. on the seminal work by Arrow on irreversible investment). Second, it can be combined with a growth objective/offensive to form an integral part of a consensus SD view – though one must be careful regarding long-term vs. short term effects, exogenous changes in preferences, and distinctions regarding preservation vs. environmental improvement [51].

The key to understanding the latter point is the following: whereas the 'strong' forms seek preservation of numerical targets – mostly fixed at present levels – for environmental variables, the preservation of the set of alternatives suggested here does not impose such restrictions. Deterioration in certain variables may be acceptable, as long as the economic benefit warrants it, and as long as there is no irreversible option-extinction, depriving future generations of the opportunity to exercise that option (e.g. rebuild a certain natural stock). Note that nothing prevents the cost of rebuilding, or restoration from being exorbitant. The actual stock left, and the ease of restoration, reflect the preferences of this generation; it is up to future generations to decide to take up the cost of restoration, if they really value it. The regrettable state of affairs that should be avoided is the impossibility of restoration, even in the face of future generations' willingness to assume the possibly exorbitant cost.

The existence of the possibility of restoration is not much consolation if the agents desiring it can not afford it: this is where the growth objective enters the scene. Growth is the irreplaceable prerequisite that will afford future generations the wherewithal to allow the exercise of restoration options; growth is necessary to prevent rendering the existence of the restoration opportunity irrelevant [52]. Hence in this light growth and sustainability will either flourish in unison, or else stagnate in discord.

References

[1]   J. Pezzey, "Sustainable Development Concepts: An Economic Analysis", World Bank Environment Paper No. 2, 1992.

[2]   H. E. Daly, "From Empty World to Full World Economics" in Goodland, R., Daly, H. and El Serafy, S. Population, Technology and Lifestyle, Island Press, Washington, DC, 1992.

[3]   C. Perrings in Models of Sustainable Development: Exclusive or Complementary Approaches to Sustainability?, Proceedings, AFCET & Universite Pantheon-Sorbonne C3E, Paris, 1994, pp. 27–43.

[4]   A. Vercelli in Models of Sustainable Development: Exclusive or Complementary Approaches to Sustainability?, Proceedings, AFCET & Universite Pantheon-Sorbonne C3E, Paris, 1994, pp. 1079–1089.

[5]   P. A. Victor, J. E. Hanna, A. Kubursi in Models of Sustainable Development: Exclusive or Complementary Approaches to Sustainability?, Proceedings, AFCET & Universite Pantheon-Sorbonne C3E, Paris, 1994, pp. 93–115.

[6]   Victor et al, op.cit. p. 95.

[7]   W. McKillop in Models of Sustainable Development: Exclusive or Complementary Approaches to Sustainability?, Proceedings, AFCET & Universite Pantheon-Sorbonne C3E, Paris, 1984, pp. 351–359.

[8]   P. Ekins, and M. Jacobs in Models of Sustainable Development: Exclusive or Complementary Approaches to Sustainability?, Proceedings, AFCET & Universite Pantheon-Sorbonne C3E, Paris, 1994, pp. 655–669.

[9]   R. Hueting, and P. R. Bosch in Models of Sustainable Development: Exclusive or Complementary Approaches to Sustainability?, Proceedings, AFCET & Universite Pantheon-Sorbonne C3E, Paris, 1994, pp. 43–57.

[10]   S. El Serafy in Models of Sustainable Development: Exclusive or Complementary Approaches to Sustainability?, Proceedings, AFCET & Universite Pantheon-Sorbonne C3E, Paris, 1994, pp. 57–73.

[11]   H. Daly, and J. B. Cobb, For the Common Good, Redirecting the Economy Toward Community, the Environment, and a Sustainable Future, Beacon Press, Boston, 1989.

[12]   P. Soederbaum in Models of Sustainable Development: Exclusive or Complementary Approaches to Sustainability?, Proceedings, AFCET & Universite Pantheon-Sorbonne C3E, Paris, 1994, pp. 1103–115.

[13]   J. M. Hartwick, American Economic Review 1977, 67, 972–974.

[14]   R. M. Solow, Scandinavian Journal of Economics 1986, 88, 141–149.

[15]   R. M. Solow (1986), op. cit.

[16]   J. Pezzey in Models of Sustainable Development: Exclusive or Complementary Approaches to Sustainability?, Proceedings, AFCET & Universite Pantheon-Sorbonne C3E, Paris, 1994, pp. 959–991.

[17]   J. L. R. Proops, and G. Atkinson in Models of Sustainable Development: Exclusive or Complementary Approaches to Sustainability?, Proceedings, AFCET & Universite Pantheon-Sorbonne C3E, Paris, 1994, pp. 819–847.

[18]   G. Marini, and P. Scaramozzino in Models of Sustainable Development: Exclusive or Complementary Approaches to Sustainability?, Proceedings, AFCET & Universite Pantheon-Sorbonne C3E, Paris, 1994, pp. 937–949.

[19]   S. Gastaldo, and L. Ragot in Models of Sustainable Development: Exclusive or Complementary Approaches to Sustainability?, Proceedings, AFCET & Universite Pantheon-Sorbonne C3E, Paris, 1994 pp. 233–245.

[20]   J. Benhaim, and P. Schembri in Models of Sustainable Development: Exclusive or Complementary Approaches to Sustainability?, Proceedings, AFCET & Universite Pantheon-Sorbonne C3E, Paris, 1994, pp. 597–613.

[21]   N. Lahaye, and D. Llerena in Models of Sustainable Development: Exclusive or Complementary Approaches to Sustainability?, Proceedings, AFCET & Universite Pantheon-Sorbonne C3E, Paris, 1994, pp. 1115–1133.

[22]   R. Kemp in Models of Sustainable Development: Exclusive or Complementary Approaches to Sustainability?, Proceedings, AFCET & Universite Pantheon-Sorbonne C3E, Paris, 1994, pp. 141–171.

[23]   C. Azar, and J. Holmberg in Models of Sustainable Development: Exclusive or Complementary Approaches to Sustainability?, Proceedings, AFCET & Universite Pantheon-Sorbonne C3E, Paris, 1994, pp. 739–751.

[24]   C. Perrings, 1994, op. cit.

[25]   G. Froger, and E. Zyla in Models of Sustainable Development: Exclusive or Complementary Approaches to Sustainability?, Proceedings, AFCET & Universite Pantheon-Sorbonne C3E, Paris, 1994, pp. 1061–1079.

[26]   A. Vercelli, 1994, op. cit.

[27]   G. Froger, and E. Zyla, 1994, op. cit.

[28]   W. Hediger in Models of Sustainable Development: Exclusive or Complementary Approaches to Sustainability?, Proceedings, AFCET & Universite Pantheon-Sorbonne C3E, Paris, 1994, pp. 317–327.

[29]   C. Azar, and J. Holmberg, 1994, op. cit.

[30]   W. R. Dubourg, and D. W. Pearce in Models of Sustainable Development: Exclusive or Complementary Approaches to Sustainability?, Proceedings, AFCET & Universite Pantheon-Sorbonne C3E, Paris, 1994, pp. 991–1005.

[31]   G. Marini, and P. Scaramozzino, 1994, op. cit.

[32]   J. G. Riley, Journal of Environmental Economics and Management 1980, 7(3), 291–307.

[33]   G. Marini, and P. Scaramozzino, 1994, op. cit.

[34]   L. Chuan-Zhong, Journal of Environmental Economics and Management 2000, 40(3), 236–250.

[35]   A. Mourmouras, Journal of Public Economics 1993, 51(2), 249–268.

[36]   J. Pezzey, 1994, op. cit.

[37]   R. B. Howarth, and R. B. Norgaard, Environmental and Resource Economics 1993, 3(4), 337–358.

[38]   J. Pezzey, 1994, op. cit.

[39]   D. Kyriakou, The IPTS Report 1999, 31, ISSN 1025-9384, can be found under www.jrc.es/iptsreport/

[40]   J. Benhaim, and P. Schembri, 1994, op. cit.

[41]   S. Speck in Models of Sustainable Development: Exclusive or Complementary Approaches to Sustainability?, Proceedings, AFCET & Universite Pantheon-Sorbonne C3E, Paris, 1994, pp. 641–655.

[42]   F. Beckenbach, and M. Pasche in Models of Sustainable Development: Exclusive or Complementary Approaches to Sustainability?, Proceedings, AFCET & Universite Pantheon-Sorbonne C3E, Paris, 1994, pp. 859–875.

[43]   A. Vercelli, 1994, op. cit.

[44]   J. Pezzey, 1994, op. cit.

[45]   Y. H. Farzin, Journal of Political Economy 1984, 92, as cited in M. E. Diedrich in Models of Sustainable Development: Exclusive or Complementary Approaches to Sustainability?, Proceedings, AFCET & Universite Pantheon-Sorbonne C3E, Paris, 1994, pp. 1044.

[46]   M. E. Diedrich in Models of Sustainable Development: Exclusive or Complementary Approaches to Sustainability?, Proceedings, AFCET & Universite Pantheon-Sorbonne C3E, Paris, 1994, pp. 1033–1045.

[47]   G. Marini, and P. Scaramozzino, 1994, op. cit.

[48]   F. Beckenbach, and M. Pasche, 1994, op. cit.

[49]   D. C. Esty, The Journal of Economic Perspectives 2001, 15(3), 113–130.

[50]   D. Stern, Journal of Environmental Economics and Management 2001, 41(2), 162–178.

[51]   M. W. Hofkes, Environmental and Resource Economics 2001, 20(1), 1–26.

[52]   D. Kyriakou, Technological Forecasting and Social Change 2002, 69 (9), 897–915.

6 A New Way of Thinking about Sustainability, Risk and Environmental Decision-Making

William E. Kastenberg*, Gloria Hauser-Kastenberg**, and David Norris***

*4103 Etcheverry Hall University of California at Berkeley, Berkeley, CA 94720, USA, kastenbe@nuc.berkeley.edu
** 1678 Shattuck Avenue 311 Berkeley, CA 94708, USA, ghauser@igc.org
*** Allmendsberg 27, 79348 Freiamt, Germany, dnorris@t-online.de

6.1 Introduction

Post-Industrial Age technologies (e.g. Biotechnology, Information Technology, Nanotechnology and Nuclear Technology) offer the promise of vast improvements in the quality of human life. And yet, as we have learned from past experience with other promising new technologies, the likelihood of undesirable consequences and unintended impacts is not negligible. The concept of sustainability is increasingly used to name this concern. We believe that the current understanding of risk analysis, which attempts to address this concern, is insufficient, because it is inconsistent with the science inherent in these newest technologies. While Industrial Age technology is based on a mechanistic and linear paradigm, Post-Industrial Age technology is not. Because the newest technologies are rooted in nonlinearity and because they have the capacity to alter life itself in unpredictable and unprecedented ways, they require a corresponding approach to risk analysis.

In this paper we begin by distinguishing the differences in context and societal impact between Industrial Age and Post-Industrial Age technologies. We then argue that such a shift in context has led necessarily to a new set of assumptions, values and beliefs that change the ethical underpinnings of technological development. And finally, we propose a new understanding of sustainability as well as an expanded approach to risk analysis and environmental decision-making, which is consistent with the new context.

Global Sustainability. Edited by P. A. Wilderer, E. D. Schroeder, H. Kopp
Copyright © 2005 WILEY-VCH Verlag GmbH & Co. KGaA, Weinheim
ISBN: 3-527-31236-6

6.2   Complicated Technology vs. Complex Technology

Because our argument rests on the premise that the newest advances in science and technology are rooted in a new paradigm, it is useful to describe here some of the fundamental differences between the science and technology of the Industrial and Post-Industrial Ages. The key distinction we draw is between systems that are "complicated" and systems that are "complex".

The context within which Industrial Age technologies are understood is based on a Newtonian/Cartesian or linear worldview where second order or nonlinear effects are (assumed) small and the boundaries are (assumed) rigid or well defined. This worldview gives rise to **complicated** systems that are characterized as atomistic (reductionism), deterministic (cause and effect) and dualistic (subject/object dualism). In other words, the properties of these systems: (1) are understandable by studying the behavior of their component parts, (2) exist independent of the observer, and (3) are only deduced from "objective" empirical observations. Some examples of such complicated technologies of the late Industrial Age include aerospace vehicles, chemical and nuclear plants, computers and robotic systems.

The context within which Post-Industrial Age Technologies are understood is based on a nonlinear worldview where second order effects are important and/or the boundaries are permeable. This worldview gives rise to **complex** systems that are characterized by at least one of the following [1]: (1) holistic/emergent – the system has properties that are exhibited only by the whole and hence cannot be described in terms of its parts, (2) chaotic – small changes in input often lead to large changes in output and/or there may be many possible outputs for a given input, and (3) subjective – some aspects of the system may not be describable by any objective means alone; that is, objectivity is considered to be only one possible way of describing system properties. Hence there may be system properties not exhibited by the parts alone[1], there may not be a causal relationship between input and output (or the output may be path dependent), and no completely analytic description for the system may be possible. Typical examples are the bottom up assembled nanoscale structures being developed for biomedical use and nanoscale chips for use in computers.

Open living systems are a class of complex systems that, in addition to the above three characteristics, possess the property of cognition. They are continually in a process of exchanging mass, energy and information with their environments. Through this exchange, open living systems seek equilibrium, either through negative feedback (returning to an existing state of equilibrium) or positive feedback (seeking a new state of equi-

---

[1] The system is simultaneously a whole and a part of a larger whole. It is often said that for complex systems, "the whole is greater than the sum of its parts". What this means is that there is an *emergent property* that is not exhibited by the parts alone.

librium). In the case of positive feedback, there also may be bifurcation points that exist far from the original equilibrium, which, when reached, lead to the attainment of a new equilibrium state at a higher level of complexity (novelty). And thus, there may be no causal relationship between the old and new states of equilibrium.

As a result of the complexity inherent in 21st Century technology, societal and environmental impacts are no longer geographically local (a bridge collapses), nor perceptible in real time (a space shuttle explodes), nor reversible (a defective automobile is recalled). Rather, complex technology can produce impacts that are geographically global (greenhouse gas emissions), imperceptible in time either manifesting very quickly (on the Internet) or very slowly (high level radioactive waste disposal), or potentially irreversible (cloning and GMOs). These impacts lead to unprecedented ethical issues such as: inter-generational risk, the threat to natural life processes and questions regarding the beginning and end points of human life. Moreover, because complex technology is evolving so much faster than complicated technology did, there is less data available now for decision-making and hence even greater ambiguity.

6.3   Errors in Thinking and Attitudes

In 1970, Gregory Bateson [2], considered by many to be one of the great social scientists of the 20th Century, wrote that there are three factors contributing to ecological damage: 1) world population growth, 2) acceleration of technological progress, and 3) certain errors in the thinking and attitudes of Western culture. He then listed seven of these errors in thinking and attitudes that emanate from the Industrial Revolution and that form the context for our current technological development:

(a) It's us *against* the environment.
(b) It's us *against* other men.
(c) It's the individual (or the individual company, or the individual nation) that matters.
(d) We *can* have unilateral control over the environment and must strive for that control.
(e) We live within an infinitely expanding "frontier".
(f) Economic determinism is common sense.
(g) Technology will do it for us.

Bateson went on to say that our best hope for the future is to change the third factor, namely, the thinking and attitudes of Western culture. Specifically, he says, "if we continue to operate in a Cartesian dualism of mind versus matter, we shall probably also continue to see the world as God versus man, elite versus people …and man versus environment. It is doubtful whether a species having *both* an advanced technology *and* this strange way of looking at its world can endure [3]."

To paraphrase Einstein's well-known observation in 1948 concerning the spread of nuclear weapons, we are faced with new questions that cannot be answered in the same framework of thinking as the one in which they were originally formulated. Recent advances in our understanding of complex systems provide a basis for such a shift in context.

6.4   The Current View of Sustainability and Risk Analysis

Currently, sustainability is usually defined as the ability to meet this generation's needs without jeopardizing the ability of future generations to meet their needs. Needs might include, for example, food, clothing, shelter, economic competitiveness, energy and national security. They also often involve the utilization of resources such as air, water, soil, minerals and other raw materials. Risk is generally defined as a characterization of the undesirable consequences and unintended impacts that can occur when a society attempts to meet its needs, often through the development of new technology. It involves measures of probability and consequence based on existing data and is usually quantified primarily in terms of public health and safety, and secondarily in terms of environmental impact. Risk can also be defined as the measure of a society's inability to meet its needs. Underlying these definitions of risk and sustainability are several articulated and unarticulated assumptions, values and beliefs that constitute an ethic.

The reductionist paradigm shapes our current understanding of sustainability and risk as well as the ethical choices derived from them. Thus, for example, risk assessment usually is reduced to a search for "causal links" or "causal chains" verified by "objective" experimental processes, i.e. by quantifying the behavior of various elements of the system [4]. The behavior of the system's elements is then integrated so as to quantify the behavior of the system as a whole [5]. While these risk assessment models are extremely sophisticated in that they are capable of evaluating a very large number of variables, the approach remains linear. Additional variables merely make the models more **complicated**; they don't, however, take account of the system's increased **complexity** due to the nonlinear nature of the interplay among the variables.

The current methodology for risk-based decision-making derives from *Utilitarianism*, the philosophy of Jeremy Bentham (1748–1832) and John Stuart Mill (1806–1873), which is predicated on the view that moral choice should be dictated by "the greatest good for the greatest number". Using cost-benefit analysis, the "good" is usually measured in terms of money. It is silent with respect to (a) the proper distribution of that good, (b) environmental justice, and (c) human and environmental flourishing. This ethical perspective is characterized by economic determinism, consistent with a linear, reductionistic worldview, and as such, is inadequate for addressing the chaotic and nonlinear nature of the issues we now face. We recognize, of course, that economics must play an important role in our environmental decision-making. However, the contextual shift we are proposing here also must include a commitment to human and ecological flourishing (discussed in next section) as well as to economics.

6.5   A New View of Sustainability

Normally, nature's life processes (biological and ecological) are self-sustaining. For example, when a change in weather pattern upsets certain of these life processes, they will attempt to adapt to the new environment, either through negative feedback (as in the case of small changes such as the seasons of the year) or through positive feedback (as in the case of large changes such as an "Ice-Age"). Alternatively, if the system cannot sustain itself in the face of the change, it will collapse. Hence, a sustainable system is one that adjusts to its new environment, either by returning to its original equilibrium (homeostasis) or by creating a new equilibrium (through chaos and bifurcation to a new and more complex level of stability).

A new understanding of sustainability must account for this dynamic property of open living systems. Natural resources, technology, world population growth and how we think about these three, are all in a positive feedback loop with each other. As the situation now stands, population growth continues, the utilization of the world's resources is accelerating, and a primary way we attempt to adapt is by creating new technology. Open living systems (including human beings) are adapting to the changing environmental landscape, and these adaptations have the effect of pushing ecological systems to a point far from equilibrium, where they eventually become chaotic or collapse.

The current definition of sustainability implies the preservation of resources for future use. The new definition we are proposing is the nurturing of the capacity for emergence in a system for future flourishing (e.g. an ecological system). This shift entails changes: (1) from a preservation of resources to a nurturing of a capacity for evolution, (2) from seeing resources as "standing-reserve" [6] to seeing resources as a requirement for human and environmental flourishing [7], and (3) from attempting to maximize human control over life processes to attempting to recognize human participation in life processes.

The notion of human flourishing was already present in the writings of Aristotle. It is a notion of the "good" as the fulfillment of what is inherent in human beings and, as such, is to be distinguished from Kant's notion of the good as what is universally "right" or from the Utilitarian view of the good as what is "useful". Flourishing is an unfolding process from within as opposed to an imposition (e.g. principles or values) from without. Furthermore, this unfolding cannot occur in isolation, but rather requires the interactive dynamic of a community, and hence, emergence is a function of co-evolution.

6.6   An Expanded View of Risk

The current notion of risk concerns the inability of future generations to meet their needs and is measured in terms of probability and consequence. Clearly we cannot maintain the status-quo and society now faces one of the following possible futures: (1) that we deplete the earth's natural resources such that biological and ecological systems can no longer sustain themselves and collapse, (2) that population growth and depletion of resources produce a new equilibrium state characterized by scarcity and environmental degradation, which diminishes the quality of life, or (3) that human activity pushes biological and ecological systems so far from equilibrium that they bifurcate into new levels of complexity with unknown consequences (good or bad) and unknowable probabilities.

Hence, a new approach to risk assessment would be based on the holism of open living systems, rather than on the reductionism inherent in the Newtonian/Cartesian paradigm. We propose the hypothesis that any evaluation of the impact of human activity on the ecology of life must shift from being based on a consideration of the sum of the individual *elements of a system* to being based on a consideration of the *emergent properties* of that system. In fact, we propose that emergent property degradation is the appropriate measure of risk for the whole of a nonlinear system, in the same way that a summative measure of risk is currently used for assessing a linear system.

6.7   An Expanded Process of Environmental Decision Making

These new and expanded definitions of risk and sustainability, in turn, become the foundation for environmental decision-making processes. Beginning with the National Research Council (NRC) study (*Understanding Risk: Informing Decisions in a Democratic Society*) [8], there has been much written about the need for analytic-deliberative processes for environmental decision-making [9–11] in multiple stakeholder situations. However, the emphasis in analytic-deliberative processes is getting the "right participants" rather than getting the "participation right." That is to say, these studies are focused on inclusivity, which is understood primarily as adding more stakeholders to the process.

Because the newest technologies are affecting core societal values as well as individual health and safety, there is a need for an expansion from merely informing the "interested and affected parties" to establishing processes for complex multi-stakeholder decision-making. However, just adding more stakeholders to the conversation merely makes decision-making more *complicated*. In fact, it thus may even decrease the chances that conflicts will be resolved, because it is a more complicated linear approach to what is inherently a nonlinear problem.

A number of deliberative processes have been suggested and some have actually been tested [9]. Suggestions range from, "*epistemological discourse*, where experts argue over factual assessment, to *reflective discourse,* where policy makers, scientists and representative members of major stakeholder groups take part, to *participatory discourse,* which is focused on ambiguity and includes legal decision making, citizen advisory groups and citizen juries [10]." But in all cases, the heart of the process must be dialogue and not merely debate. A debate leads to either a triumph of one of the initial points of view over the others, or perhaps to a compromise. Dialogue, on the other hand, can lead to the emergence of a new possibility that was unthinkable prior to the start of the dialogue. Dialogue is distinguished here from its usual meaning [12]. In fact, it is nonlinearity in action [13]. When in a process of dialogue, prior assumptions, values and beliefs are suspended, knowing is replaced with finding out, known answers replaced with new questions, winning or losing with cooperating, power with respect, and proving points with exploring possibilities.

6.8   Conclusions

Risk analysis, to date, has been used primarily to address the potential societal and environmental impacts of complicated systems. It was developed by the U. S. Space Program in the 1950's and 60's with the advent of Failure Modes and Effects Analysis in an attempt to both understand and correct missile and rocket launch failures. Risk assessment came of age with the publication of the Reactor Safety Study (WASH-1400) in 1975, but only after 75 or so nuclear power plants already had been designed, built and operated in the U. S. Risk management only came to prominence after the accident at Three Mile Island-Unit 2 in 1979.

We are now challenging the field of risk analysis to address issues concerning sustainability with respect to complex systems before they are widely deployed and before irreversible consequences have occurred. It is a challenge not to be taken lightly.

References

[1]   Science 1999, 284(5411), 79–109.
[2]   G. Bateson, Steps to an Ecology of Mind, The University of Chicago Press, Chicago, 1999, p. 500.
[3]   G. Bateson, Steps to an Ecology of Mind, The University of Chicago Press, Chicago, 1999, p. 337.
[4]   B. Wynne, Global Environmental Change 1992, 2, 111–127.
[5]   S. Kaplan, and B. J. Garrick, Risk Analysis 1981, 1(1), 11–27.
[6]   M. Heidegger in The Question Concerning Technology and Other Essays, translated by William Lovitt, Harper and Rowe, New York, 1997.
[7]   Proceedings of a Workshop on Ethics and the Impact of Technology on Society, can be found under http://www.nuc.berkeley.edu/html/research/ethics/index.htm, 2003.
[8]   National Research Council, Understanding Risk: Informing Decisions in a Democratic Society, 1996.
[9]   G. E. Apostolakis, and S. Pickett, Risk Analysis 1998, 18(5).
[10]   O. Renn, Environmental Science and Technology 1999, 33(18), 3049–3055.
[11]   S. Tuler, and T. Webler, Risk, Health, Safety and Environment 1999, 10, 65–87.
[12]   D. Bohm, On Dialogue, Routledge, 1996.
[13]   G.-H. Kastenberg, W. E. Kastenberg, D. Norris, Science and Engineering Ethics 2003, 9.

7   Humility and Establishing the Sustainable Environment

Edward D. Schroeder

Department of Civil & Environmental Engineering, University of California,
1 Shields Avenue, Davis, CA 95616, USA, edschroeder@ucdavis.edu

7.1   Introduction

The concept of environmental sustainability has gradually grown over the past 50 years as a result of greatly increased knowledge and understanding of wildlife ecology, visible and measurable changes to our environment, and the development of the worldwide environmental movement. The necessity of moving toward a sustainable environment is exceeded in importance only by the need to define sustainability. Human societies have taken on many great causes since the beginning of industrialization and very often the results have been much different than anticipated. Based on the history of human intervention into very complex issues a great deal of humility should be incorporated into our attempts to develop a sustainable environment. An example is provided by one of these causes, arid regions water management, that is closely tied to developing sustainable environments

Arid regions, or deserts, are usually characterized by annual precipitation less than 25 cm. Semi-arid or semi-desert regions have annual precipitation up to 50 cm. In both cases, annual precipitation varies considerably from year to year. Water management in arid and semi-arid areas is an ancient endeavor. Efforts to manage water are recorded in the early histories of Egypt, Mesopotamia, and China and include many examples of development of huge water systems and failure of these systems due to a lack of understanding of the complexities of the natural environment. The programs developed in California over the past 100 years provide some of the most poignant examples of the need for humility in environmental management.

California is composed of 10 distinct drainage basins, as indicated in Figure 7.1. Two of the basins, the South Lahontan and the Colorado River, which includes the Mohave Desert, Death Valley, and the Imperial Valley, are principally desert and comprise approximately one-quarter of the land area of the state. The North Lahontan Basin, a narrow strip along the north east of the state that includes Lake Tahoe is also semi-arid and drains to Nevada. Two of the basins are semi-arid coastal plains extending approxi-

Global Sustainability. Edited by P. A. Wilderer, E. D. Schroeder, H. Kopp
Copyright © 2005 WILEY-VCH Verlag GmbH & Co. KGaA, Weinheim
ISBN: 3-527-31236-6

mately 800 km between San Francisco and the Mexican border and approximately 100 km inland and vary in annual rainfall from 15 to 50 cm.

Two basins, the San Francisco Bay and North Coast, comprising about 10 percent of the land area of California, receive relatively large amounts of precipitation, particularly near the coast, with some localities in the North Coast basin averaging over 200 cm per

Figure 7.1 California drainage basins and typical ranges of annual precipitation. Three of the basins, the Sacrament River, San Joaquin River, and Tulare Lake Basin form the Great Central Valley. Over 80 percent of Californians live along the coast in the San Francisco, Central Coast and South Coast drainage basins. Land area of California is 411,000 m^2, about three times that of England, 15 percent larger than Germany and 75 percent that of France

year. This region is the only area of the state that is not largely arid or semi-arid. Thirty million of California's 35 million people live in the narrow strip extending from the Mexican border through the San Francisco Bay drainage basin. Remaining is the Great Central Valley, 750 km in length, 150 km in width, bounded on the east by the Sierra Nevada and on the west by the Coast Range and comprising approximately forty percent of California's land area. Although home to five million people, the principal industry of the Central Valley is agriculture and the southern portion is the most productive agricultural region in the United States. The valley floor is arid to semi-arid, with precipitation ranging from 12 to 100 cm per year, but the western slopes of the Sierra Nevada receive large amounts of snow, often 250 to 500 cm. Throughout California nearly all (85 to 95 percent) of the precipitation occurs between November and April. In essence the state has a six-month drought every year. Moreover, most of the precipitation is confined to a few storms during the wet season, as shown in Figures 7.2A and 7.2B.

Water management plans for California were developed before significant consideration was given to water quality and long before there was significant understanding of the complexity of aquatic ecosystems. Moreover, the history of irrigated agriculture, which accounts for 80 percent of all water use in California, was given little consideration, if not ignored. Reservoirs and canals were constructed to store water for use during the dry season and to transport it for application in the Central Valley and the coastal plains. Virtually all of the rivers in California are controlled and for most of the year flow in rivers and streams in the Los Angeles, Santa Ana, and San Diego watersheds (commonly known as Southern California or SoCal) is principally irrigation runoff and treated wastewater. Approximately 7.5×10^9 m^3 per year of snow melt are moved from areas north of San Francisco to the southern Central Valley and the coastal plains each year using a system of reservoirs, canals and pumping stations. The system was constructed with the belief that fresh water discharged to the ocean was wasted and that the reservoirs would provide the added benefit of flood control. Implementation of the California Water Plan [1] has resulted in a number of surprises.

One surprise was loss of spawning beds due to trapping gravels behind dams and stopping the floods that washed out sediment. Winter run Chinook Salmon in the Sacramento River decreased from over 40,000 in 1970 to 500 in 1989 due to degradation of spawning, rearing, and migration habitats [2, 3, 4]. Another surprise was the impact of salts and chemical applications associated with the intense agriculture supported by the transported water with the two major reservoir and canal systems constructed to store snow melt and transfer it to farms in the Central Valley and the metropolitan areas of Southern California. Pesticides applied to irrigated land have contaminated groundwater aquifers and resulted in the loss of a large number of drinking water wells. Disposal of agricultural drainage has been addressed through only very stop-gap measures, one of which was Kesterton Reservoir where selenium contained in agricultural drainage resulted in severe birth defect and reproduction problems in resident bird populations [5]. Still another surprise was the loss of sand from coastal beaches due to trapping eroded materials upstream and paving over of coastal areas in the south coast region [6]. Most of the wetland areas along the Sacramento and San Joaquin Rivers and along the coast have been drained with resulting loss of wildlife habitat and feeding grounds for migratory birds.

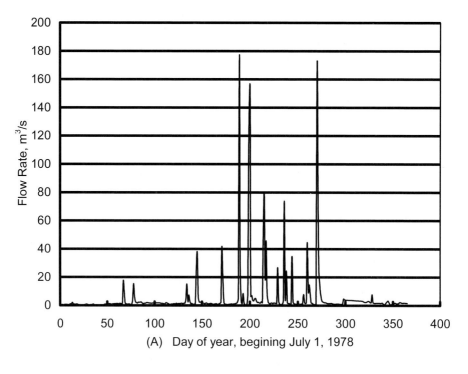

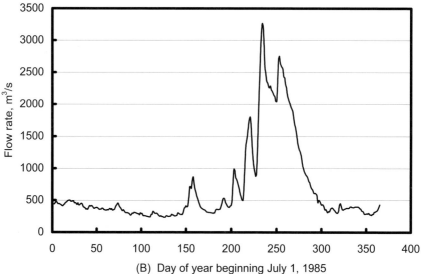

Figure 7.2 Characteristic flow patterns in California Rivers
A. Flow in the Los Angeles River, Los Angeles California in the South Coast Basin during 1978–1979 hydrologic year
B. Flow in Sacramento River in the 1985–1986 hydrologic year, one of the wettest on record

agricultural region in the United States and value of metropolitan areas along the California coast. The question of how we would develop our water system if we could begin again is only now being considered. However, this question should be asked and carefully evaluated using a broad spectrum of objectives. What would California be like today if our current understanding of the physical, chemical, and biological world were available in 1850 or 1900? What mistakes are we making today because of a lack of knowledge and how are we to account for this lack of knowledge in our planning?

7.2   Pristine Environments and Preservation

The goal of a sustainable environment must be referenced to measurable values that are accepted by ordinary people, by industry, and by governments. At present there is a tendency to use an idealized pristine environment free of all contamination as a model and as an objective. However, even environments untouched by humans will have measurable pollution. Animal faeces, dead animals and decaying plants, will contaminate water and soil. Pathogens, such as *Salmonella* spp. and *Giardia lamblia* will exist in such environments even if humans are not present (7). Metals will leach from mineral formations and in many cases concentrations may exceed health, or more likely aquatic habitat standards.

The pristine environment concept is most easily objectified in areas that have received relatively little human impact such as the Antarctic, the Amazon rainforest, and Alaska. In areas such as the east coast of the United States and most of Europe there is little left that might be thought of as truly natural or native habitat. How would we define a pristine environment objective for Germany where essentially all of the land has been managed for several hundred years? The problem must be restated in such a way that the focus is on creating a definition of a desirable environment and having a clear rationale for the definition. Such definitions are particularly important in discussing water quality because when most people think of natural water the immediate picture is a clear, cool, shaded, gravel bottomed stream. The facts of rotting vegetation, animal wastes and carcasses, and insects are rarely included.

Coupled to the pristine environment concept is the desire for preservation – to keep things the same. In our little university town of Davis letters to the editor of our newspaper often have statements such as "We moved here two years ago to live in a small community and now we find growth is out of control. The City must stop issuing building permits". Another common letter type reflects a desire to be close to nature – "We have lived in our newly constructed house for a year and have enjoyed watching the burrowing owls on the vacant land next door. The City must change the intended use of this land from playing fields to natural habitat so that the owls will be preserved". Such comments are illustrative of how even well educated people approach large scale issues. In the case of Davis, people move here for logical reasons but fail to recognize their own impact on the system. On a larger scale, efforts to preserve old growth forests, wildlife habitat or agricultural land (which once was habitat) are increasingly in conflict with the desirability of old growth lumber, recreational development, and the growth of cities.

Preservation of species often requires large tracts of contiguous habitat and the maintenance of a complex ecosystem that may be inconsistent with urban society. For example, the cassowary, a large flightless bird living in rainforest habitat of northern Queensland in Australia and parts of New Guinea, has greatly declined due to loss of habitat and increased use of habitat by humans (particularly humans with dogs). There are perhaps 3000 cassowaries left at present. Because seeds of fleshy fruit, the primary food source for cassowaries, are excreted undamaged, the bird plays an important role in dispersal of seeds in the rainforest. In particular, the cassowary plum produces a seed containing toxic alkaloids that only the cassowary can digest. Thus the bird is essential for the dispersal of the seeds and the survival of the cassowary plum. When and if cassowaries become extinct the tree will also become extinct. The cassowary plum undoubtedly interacts with other animals and plants in the rainforest in providing the unique habitat that exists at present but of which we have little understanding. Quite possibly the loss of the cassowary plum would result in the loss of other plants and animals in something of a domino effect. We can assume that our answer to the question of preservation of the cassowary plum will be permanent if the bird is allowed to go extinct. Preservation of both the bird and its habitat may require sealing off a relatively large area of rainforest from development and possibly from use for recreation.

The cassowary presents a relatively simple case because it is located in areas that have only recently come into significant human intervention. Rainforest could be cordoned off with restricted entrance requirements without expending extremely large sums of money if governments wished to do so. Such an approach would be a departure from conventional park development because the general public would, to a great extent, be excluded. Quite likely the approach would be controversial because continued public expenditure would be required for monitoring, research, and maintenance and there would not be a clear public benefit. Arguments that maintaining diverse gene pools is important, that environmental preserves will benefit future generations, or that human society has a moral obligation to maintain each and every species may not be politically viable.

Preservation programs in the United States have included efforts to bring predators such as mountain lions, bob cats, wolves, and bears back into former habitats for these animals have met with mixed success and mixed acceptance. Proponents cite the desirability of maintaining wild populations of these animals because of their beauty, their role in the ecosystem, the need to maintain genetic diversity, the need to preserve the natural environment, and the rights of the animals themselves. Predator species were often the subject of eradication programs because of the desire to protect cattle, sheep, and other domestic animals. Such programs existed as recently as 1980 in some localities of the United States. Increased understanding of the interaction of predator species with their prey and with the broader environment has resulted in the development of new programs, particularly in national parks and remote areas of the Rocky Mountains, Cascade Mountains and the Sierra Nevada.

Mountain lions (pumas) had a natural range covering most of the Americas below 3,000 m elevation. Through loss of habitat and eradication the animals are now limited to Mexico, Central America, the western United States and far-western Canada. They have proven adaptable to suburban environments and where urban growth has expanded

into suitable habitat, such as the Sierra Nevada foothills, the cats have substituted domestic animals and on three recent occasions joggers and bicyclists for their natural prey,
although only a dozen humans have been killed over the past 100 years. California is
believed to have approximately 5000 mountain lions, enough to make an important contribution to the ecosystem through thinning of deer, elk, and smaller mammal populations. Wolves were completely eradicated from the contiguous 48 states, in part because
as their natural prey decreased due to expanding agriculture the animals substituted domestic animals as a food supply. Wolves were reintroduced to Yellowstone Park and
parts of Idaho and Montana in 1995. Prey has been plentiful, few domestic animals have
been killed, and there is a program for reimbursement of ranchers when a kill takes
place. The predator community populations are shifting, with the wolves pushing aside
coyotes to an extent. Smaller mammals that are the chief prey of coyotes are becoming
more common. Additionally, the antelope, deer and elk herds are becoming more stable
because weaker animals are taken by wolves rather than continuing on until an extremely hard winter decimates the population. The somewhat smaller herds of herbivores
are resulting in better vegetation as well.

Grizzly bears are large (200 to 600 kg), unafraid of humans, and once ranged through
most of the western United States, Canada, and Alaska. Eradication programs in the
United States reduced the population from an estimated 50 to 100,000 to less than 1000
today located in Yellowstone Park and along the US-Canadian border. Grizzly bear
incidents with backpackers, hikers, and campers occasionally occur, often as the result of
mistakes by humans in managing food. A more significant problem are bears that become habituated to human food by consuming garbage, food left at campsites, and food
left out at rural homes. Once habituated, it has proven extremely difficult to keep them
away from sites where human contact might occur and killing is the solution. In California, black bears remain in the high Sierra Nevada and along the northern coast. Interactions with backpackers occur every summer, although injuries are rare. Proponents and
opponents of reintroduction of these animals into areas now occupied by humans are not
finding a great deal of common ground in their arguments.

A corresponding issue is the program to rid California's Channel Islands, near Los
Angeles, of non- native black rats that threaten native sea bird populations. The U. S.
Park Service has been attempting to kill the rats with poison. Rob Puddicombe, of Santa
Barbara was arrested in October, 2001 for spreading pellets containing vitamin K, an
antidote for the rat poison. He was quoted in the Sacramento Bee (December 16, 2002)
that "park rangers have no right to choose which animals should live and which should
die. It's a topsy-turvy world when poisoning wildlife from helicopters is good and feeding wildlife is a crime. As far as I am concerned, this is like ethnic cleansing; it's a jihad
against non-native species."

In considering preservation we need to ask if we must preserve all species everywhere. At one time an estimated 60,000,000 bison roamed the central plains of the
United States in huge herds. The manner in which the American bison was brought to
near extinction is not a pretty story [8]. However, if the herds of bison had been preserved the agricultural bread-basket of the country would be quite different today. Thus
there is a question of placing value on species and resources that are essentially irreplaceable. Is maintenance of a few small herds of bison that no longer interact with or

impact their natural habitat satisfactory? Should we attempt to return major areas of land to the conditions present before the arrival of Europeans and for what reason should this be done? These questions take us back to the Los Angeles River (see hydrograph of Figure 2A). In 1800, flow in the Los Angeles River was most likely quite similar to that of today in a very general sense. The river was nearly dry most of the time and served as a drain for the 15 to 20 storms that passed through the basin each year. The hydrographs for each storm would have been much broader and would have lower peaks because of retardation by natural formations and infiltration into the soil. Paving over of much of the basin has resulted in much more rapid transport of runoff to the river channel and a significant decrease in infiltration. Should California set an objective of returning the basin to the conditions of 1800 (or some other arbitrary date)? What impact would such a return have on life in the basin, which now includes approximately 20 million people and one of the major industrial centers of the world?

7.3   Specieism

Many people have concluded that humans do/should not have greater rights than other species. Animal rights proponents such as Stephen Best, Chair of the Department of Philosophy at the University of Texas at El Paso [Letters, *Alantic Monthly*, March, 2003] argues that humans are simply one species that has evolved and that other species (for some this includes trees and perhaps lower forms) have the same *natural* rights. This train of thought takes us well beyond deep concern with the morality of factory farms, genetic modification, growth hormones, and protection of old growth forests. There is an assumption of similar or equal cognitive pain among the various animal species resulting from the living environment, being raised to be eaten, the slaughtering process, and use in experiments and chemical testing. The argument for plants is analogous and activists have chained themselves to trees, lived high in their canopies for months to years, and given trees names to prevent cutting. There is good reason to ponder the issues because modern animal production has developed on a purely economic basis and welfare of the animals has been given little consideration. However, general acceptance that other species have equal rights to our own species would certainly raise questions about the definition of a sustainable ecosystem. The argument of Mr. Puddicombe that non-native species have a right to be on the Channel Islands needs to be addressed in this context, also. Non-native species often arrive in new territories as "stowaways" on ships, airplanes, and trucks and their arrival causes severe disturbances in evolved ecosystems. If Mr. Puddicombe's beliefs attain wide support "immigration controls" on plants and animals will be impossible and many species will be eliminated through a somewhat artificially introduced competition. A key issue is reaching a consensus about the relative ranking of species. If such a consensus is not reached by the intellectual community (including the religious community), governments, and the general public, the decision making process required to develop a sustainable environment will be very difficult. Attaining a consensus will not be easy. Animal and plant rights activists, including such organizations as the Animal Liberation Front and Earth First, and believers in laissez faire economics provide a defining envelope within which most approaches to environmental management can be placed. The two extremes are important because they raise many of the most difficult questions about environmental management. Unfortunately, adherents to the boundary positions tend to hold their views as absolutes and find compromise extremely difficult, if not immoral. Thus, although their positions must be carefully considered and will in many cases become important aspects of management plans, it is unlikely individuals or groups at the edge will contribute to maintaining sustainable programs.

7.4   Where Do We Go/How Do We Get There?

Human population and human activity have increased to an extent that we are significantly impacting our environment and the general trend is a rapid rate of degradation. However, we should also note that the large majority of people are living much better, and for a longer time, than ever before. Many believe that the ecosystem we rely upon is becoming unstable and that cataclysmic events, perhaps new plagues, rather than a slow decline may occur. Optimists believe that we can deal with any problem through science and planning. The planning aspect is perplexing because our past attempts should make us very humble indeed. However, like each new generation of adolescents who assume they have a much clearer view of the world than their parents, we tend to assume that at last we really understand the complexity of our environment.

The answer would seem to be in developing a consensus about the definition of sustainable ecosystem and adopting flexible and correctable mechanisms for achieving it. Our understanding of the interactions in the biosphere, the impacts of physical and chemical factors on the ecosystem, and the impacts of human activities on the physical and biological world is increasing rapidly. Mathematical models of the interactions are improving as is our ability to apply models in environmental decision making. However, humility must remain in our process.

In 1985 cryptosporidiosis was not considered a waterborne disease and was not mentioned in any textbook on water quality. In 1993 over 400,000 people in Milwaukee, Wisconsin were infected by *Cryptosporidum* through a contaminated water supply – the worst outbreak of waterborne disease in recorded history.

We can expect additional surprises despite application of our best science and planning. Thus the need for flexibility in planning and decision making is very great. We have a history of protecting "solutions" and plans that have been encoded into laws and regulations because change presents temporary problems. Developing an ability to modify, or perhaps completely change, plans in which much has been invested will be essential to success. Humility will be a key factor in finding a path to a sustainable ecosystem – assuming we can reach a consensus on what that system is.

Perhaps, a return to the Los Angeles drainage basin and the hydrographs of Figure 2A will provide a modest model of the approach. The storm hydrographs shown in Figure 2A result from paving of large amounts of land and channelization of the Los Angeles River, which is really a paved canal for most of its length today. The reason the river was channelized was flood control and the flooding problems resulted from the greatly increase amount of impervious land in the drainage basin. A higher fraction of precipitation becomes runoff when land is paved and the concurrent result is a decrease in water percolating to the groundwater aquifer. Groundwaters of the Los Angeles basin have long been mined. For nearly 50 years there has been a program to maintain the groundwater aquifer and minimize sea water intrusion through the infiltration of highly treated municipal wastewater at various points in Los Angeles and Orange Counties. Increasing the fraction of storm water percolated to aquifers would result in improved groundwater quality, decreased flood flow peaks in the Los Angeles River, and perhaps a decreased need for wastewater reuse for domestic purposes. Portions of the Los Angeles River might be returned to a more natural state and habitat for native species of plants and animals might be improved. The question is now focused on the mechanisms used to increase percolation. Factors to consider include the need to minimize standing water (West Nile Virus has recently become a serious factor in the region and mosquito control will be a facet of any plan), and lack of appropriate sites for reservoirs. Making preservation and increasing permeable surface a factor in development, such as is done in Florida, Germany, and elsewhere may be a viable approach. Parking lots, sidewalks and similar public areas could be retrofitted. However, methods for capturing runoff from roof drainage in established areas, including urban centers would surely be necessary for a significant increase in the percolating fraction to be established. Because the quantities from individual buildings and houses will be small for each storm, it may be possible to use vertical percolation systems (reverse wells) that can be constructed using conventional boring technology. Surprising to many, stormwater is often highly contaminated and treatment may be required. The use of vertical treatment systems with layers of zeolite or activated carbon to remove metals and organics has been suggested [9]. Implementation of such a plan would require a great deal of study and evaluation of costs and benefits. However, the elimination of the concrete scar across the basin that is the Los Angeles River, the possible development of habitat and return of native species of plants and animals (bears and mountain lions would seem inappropriate), the increase in green area, and the improvement of groundwater quality and possibly quantity may be enough to induce further investigation. Because of our increased understanding of the ecosystems and the native habitat, and the availability of mathematical models to describe impacts on the aquifers and the ecosystems a better method of managing stormwater in the Los Angeles Basin may be possible. However, such a project would be so large that all of the impacts and interactions cannot be foreseen. Humility and the expectation of error must be implicit in such a plan or the final result will be a different version of the current situation. The definition of sustainability remains in question because the "revised" environment would in no way be an approximation of what would have evolved without human intervention.

7.5 Conclusions

Human society has a record as long as history of developing large projects with little understanding of the impact that will be made on the environment. Until recently, the lack of knowledge about the earth's ecosystems and our inability to measure impacts provided an excuse for our actions. However, we are now able to observe many of the impacts and we understand that complex interactions exist throughout the physical, chemical, and biological world. In attempting to develop a sustainable society there must be great care to ensure the process and product do not further damage the environment. Great projects of the 20th century, designed to alleviate social problems, provide energy, improve nutrition and food supplies, and provide faster transportation and communication were often successful in terms of the narrow objective and disasters in terms of the larger human environment.

Plans for a sustainable environment must incorporate a type of correctional algorithms analogous to the Kalman filters used in process control to make course corrections during application. Human experience with planning allows no reason to suspect that programs for a sustainable environment can be placed on autopilot or that plans will not need to change substantially as time progresses and conditions change.

There is not a generally accepted view about the rules for environmental sustainability. The most aggressive activists for environmental causes often have narrow interests and little flexibility with respect to ethical and moral decision making. Society will need to develop a more broadly accepted definition of the role of humans versus other living things than exists at present if programs for a sustainable world are to be successful.

Acknowledgment

The thoughtful comments and suggestions of Professor Stefan Wuertz of the University of California, Davis were very helpful in writing the manuscript and are greatly appreciated.

References

[1]   California Department of Water Resources, The California Water Plan, California Department of Water Resources, Bulletin 3, Sacramento, CA, 1957.

[2]   National Marine Fisheries Service, Southwest Region, Biological Assemssment for The Fishery Management Plan for Commercial and Recreational Salmon Fisheries off the Coasts of Washington, Oregon, and California as it affects the Sacramento River Winter Chinook Salmon, Long Beach, CA, 1996.

[3]   L. W. Bradford, and J. G. Brittnacher, Conservation Biology 1998, 12, pp. 65–79.

[4]   T. J. Mills, D. McEwan, M. R. Jennings, in Pacific almon and Their Ecosystem (Eds. D. J. Stouder, P. A. Bisson, R. J. Naiman), Chapman & Hall, New York, 1997.

[5]   H. M. Ohlendorf, and R. I. Hothem in Handbook of Ecotoxicology (Ed. D. J. Hoffman), Lewis Publishers, Boca Raton, FL, 1995, pp. 577–585.

[6]   N. Pan, and E. D. Schroeder, Beneficial aspects of stormwater, Report submitted to the Division of Environmental Analysis, California Department of Transportation, Interagency Agreement no. 43A0073, Center for Environmental & Water Resources Engineering, University of California, Davis, 1998.

[7]   E. D. Schroeder, W. M. Stallard, D. E. Thompson, F. J. Loge, M. A. Deshusses, H. H. J. Cox, Management of pathogens associated with storm drain discharge, Report submitted to the Division of Environmental Analysis, California Department of Transportation, Interagency Agreement no. 43A0073, Center for Environmental & Water Resources Engineering, University of California, Davis, 2002.

[8]   M. S. Garretson, The American Bison 1938, New York Zoologocial Society, New York.

[9]   P. A. Wilderer, Personal communication, Schliersee, Germany, 2003.

8 Conflicts and Conflict-solving as Chances to Make the Concept of Sustainable Development Work

Armin Grunwald

Institute for Technology Assessment and Systems Analysis (ITAS), Research Centre
Karlsruhe, Hermann-von-Helmholtz-Platz 1, 76344 Eggenstein-Leopoldshafen,
Germany, grunwald@itas.fzk.de

8.1 Sustainability as a Conflict-generating Vision

Sustainability as a societal vision is, on the one hand – at least on the political-programmatic level – not only potentially acceptable, but does, in fact, meet with correspondingly broad approval across all societal groups and political positions, nationally and internationally. The number of nations which have signed and ratified the documents of Rio 1992 and the corresponding follow-up papers and the numerous local or regional activities are impressive.

On the other hand, sustainability's conflict potential can't be overlooked. As soon as relatively concrete goals or even strategies of societal action for attaining sustainability are put on the agenda – at the latest – it becomes obvious that the usual antagonistic societal values and interests are lurking behind the programmatic consensus.

Due to this fact, the opinion is often expressed that sustainability is a concept without content, or that sustainability is a harmonistic wrapper (meant here in an analogous sense to that of the definition in the field of biblical hermeneutics; the harmonistic tradition smoothes the disparities in the biblical text in a manner that imposes greater strain on faith than do the disparities themselves, i.e.: sustainability is, according to this objection, supposed to create a harmoniousness which in reality doesn't exist) over heterogeneous and incompatible goals, and can therefore only have rhetorical functions. It has sunken, so some people argue, to the level of arbitrariness, and no longer has any power to "make a difference". In order to refute these contentions, it would be necessary, on the one hand, to make clear what the concept of sustainability comprises and what not. On the other hand, the principle of sustainability should not appear to be merely a plaything of conflicting interests, but has to demonstrate and realize possibilities for settling such conflicts in a constructive and, optimally, in a "sustainable" manner.

Global Sustainability. Edited by P. A. Wilderer, E. D. Schroeder, H. Kopp
Copyright © 2005 WILEY-VCH Verlag GmbH & Co. KGaA, Weinheim
ISBN: 3-527-31236-6

The objection that nothing more than harmonistic meaninglessness is hidden behind the concept of sustainability, that talking about sustainability would therefore either be of no consequence, or could be arbitrarily instrumentalized or misused, can be interpreted in a number of ways which allow a better understanding of that objection:

a) Sustainability as mere design: The postulate of sustainability has, in this version of the objection, no content. Of course, nobody can be opposed to a person's pursuing his or her economic interests in a sustainable manner which "satisfies the needs of the present generation without compromising the ability of future generations to satisfy their own needs"[1], but acceptance of this understanding says nothing specific regarding content. People who can all generally agree with this statement can still compete further for diametrically opposed aims.

b) Sustainability as an ideological illusion: The concept of sustainability conceals in this manner the conflicts of interests among the real actors and the actual power constellations. It is instrumentalized as an ostensible legitimisation of power and of particular vested interests, for instance, in the question of the relationship between securing continued affluence in the industrialized nations and the perspectives for the developing countries. The danger is that each social actor or group may define its "own" sustainability – the farmers, the industry, social movements, political parties, authorities or others. All of them could then claim to promote sustainable development but with using diverging or contradictory understandings of sustainability.

c) Sustainability as a utopian hope: A further point of critique is that the concept of sustainability is overtaxed, in any case, whenever more than ecological sustainability only is subsumed under it [2]. If sustainability should be used as a collective designation for everything "noble, helpful, and good", then this would be impracticable, could lead to arbitrary conclusions, and would arouse expectations which can't be fulfilled. The concept of sustainability as an integrative-utopian aspiration is, according to this contention, a harmonistic illusion which blocks the view onto the real problems.

These doubts have to be taken seriously – at least for the time being. A theory of sustainability which lays claim to relevance in practice and to presenting contributions toward solving societal problems has to make clear how it reacts to these objections. The concept, theory, and operationalization of sustainability have to fulfil at least the following requirements in order to avoid slipping into arbitrariness:

- Specification of the fields of application: exactly what they apply to, and what not must be stated, i.e., a field of responsibility for sustainability must be delimited from questions for which sustainability isn't relevant.
- Unambiguous judgements: distinctions must be made between "sustainable", and "unsustainable" or "less sustainable" possibilities within the areas concerned.
- Operationalizability: concrete ascription must be made of these judgements (sustainable/unsustainable) to societal circumstances or possible developments (e. g., by means of indicators to be derived from the concept and theory of sustainability but also by using explanatory cause/effect knowledge about the developments under consideration).

It turns out that conflicts over sustainability don't first arise – as has often been maintained – at the point when concrete measures are discussed. Rather conflicts are unavoidable as early as on the conceptual level where the basic understanding of sustainable development has to be clarified. At least the following types of conflict can break out at this stage:

1. Conflicts of Demarcation: What belongs to the subject area that the principle of sustainability should be applied to, and what doesn't? Is it solely a matter of responsibility for the future or of distributive justice in the present? Is it a question of conservation, or of development? What should be protected or developed under the banner of sustainability? Which legitimate societal goals are there beyond or outside of the concept of sustainability? In the integrative concept of sustainability [3, p. 172], the field of application is defined, for example, so, that the sustainability rules are formulated not as the sum total of all desirable societal goals, but as minimum requirements for a lastingly humane existence. Societal goals which extend above and beyond this level no longer belong to sustainability's areas of study and evaluation. The exact determination of this line of demarcation is, however, obviously connected with societal conflicts, because it is neither on the national nor on the global level at all clear what should be included in these minimum requirements, what the attribute "humane" includes, and what it excludes.

2. Conflicts about the substitutability of different parts of the overall societal capital: Every generation disposes over a certain productive potential, which is made up of various factors (natural capital, real capital, human capital, knowledge capital). Sustainable development demands in general, that the stock of capital which exists within a generation be handed down as undiminished as possible to future generations – whereby, however, two fundamentally different alternatives are conceivable (cf. [4], p. 110 ff.). On the one hand, one could stipulate that the sum of natural and human-made capital be constant in the sense of an economy-wide total; on the other hand, one could require that every single component of itself has to be preserved intact. The former path is sensible if one assumes that natural and human-made parts of the overall capital are completely interchangeable (weak sustainability). The latter path is advisable if one assumes that human-made and natural capital stand in a complementary relationship to one another (strong sustainability). The controversy over both of these strategies, that is, over the question, how the heritage which is to be handed down to future generations should be composed, is one of the central problems of the sustainability debate (cf. [5]). There are also intermediate positions, sometimes designated as "sensible sustainability"[6]. Due to this approach, the substitution of natural capital by human-made capital is held to be admissible to a limited extent, as long as nature's basic functions (the immaterial ones as well) are maintained.

3. Conflicts over Priorities: Whenever it is a question of the mutual relations of the various (ecological, economic, social, and political) dimensions of sustainability, or of the relation between inter- and intragenerational equity, careful consideration and weighing of priorities are imperative. The proposed approaches to sustainability in the various dimensions will not always mutually reinforce each other and lead to

"win-win"-situations. For instance, the precept of conserving landscapes of a particularly characteristic nature and beauty can come into conflict with the need for securing an independent livelihood, as far as the local population is concerned – a classical conflict in environmental conservation policy. It is then necessary to weigh up goals and values and to set priorities which, as a rule, quite obviously give rise to societal conflicts.

4. Conflicts over the Choice of Indicators: Appropriate and meaningful indicators for sustainability can't be derived logically and deductively from the sustainability rules. Rather, different indicators are conceivable, which respectively set different accents. The determination of indicators influences further questions, such as, which parameters should be chosen for long-term observation, or for which parameters targets should be set and commitments be made. Because the choice of indicators is, therefore, not value-neutral, it can be fraught with conflicts – and often is, as the pertinent discussion shows. These conflict levels make clear that conflicts over sustainability not only occur, as is often discussed, on the strategic level of concrete measures and their realization, but that they are inherent in the very conceptualization of sustainability. In addition, the usual conflicts arise in the further strategic operationalization of sustainability whenever it is a question of specific measures or their consequences. The distribution of the burdens and risks of measures for promoting sustainability is, as a matter of course, conflict-laden in a pluralistic society.

5. Conflicts of Distribution: Further conflict potentials can arise on the strategic level when it comes to translating the principle of sustainable development into concrete responsibilities of action for societal actors. When, for example, one has to decide which contribution the transportation industry and which the power supply industry should bring toward realizing a national CO_2-reduction target. When the contributions of various nations to common goals are set, quite substantial conflicts of interests flare up (as could be observed, for instance, in the Kyoto follow-up conferences). On the one hand, conflicts of distribution arise because of the winner-looser problems, and on the other hand, due to the finiteness of scarce resources, such as drinking water or soil.

These different types of sustainability conflicts mentioned have their origin in the diversity of the conflicts inherent in a pluralistic society (as, for example, differing conceptions of justice, of responsibility, of the role of the welfare state, or of the economic system). Different and contradicting interests between social actors, between NGOs and industry, between political parties or between developed and developing countries are leading to such conflicts. These conflicts aren't just dissolved into thin air by their common relevance for sustainability, but come into play again when sustainability is to be made operable. This is in no way surprising.

Of greater interest with respect to this paper is the fact that not only already existing societal conflicts play a role, but that the imperative of sustainability itself is also the source of additional conflicts. As soon as the question of justice – and this is the essence of sustainability – is extended beyond the small national or regional circle of the present generation to the global scale and to future generations, completely new questions and additional distributive problems arise – with the corresponding lines of conflict. In this category belong questions of the sort, whether and how much abstinence can be expected of those presently living (in the Western nations) in the interest of future generations, and if so, how this abstinence should be distributed among and within nations. This situation is the clearest proof of the fact that the principle of sustainability is anything but harmonistic, and can even be the origin of conflicts.

It is the extension of the time and space dimension inherent to the imperative of sustainable development which leads to new types of challenges in the reflections on justice and equity. Conflicts between the assumed needs and interests of future generations – obviously, there is already a problem of knowing enough about them – and the interests of people living today arise. Why should we renounce on realising certain needs in favour of future generations which we will never meet? The global dimension of sustainability [7] leads to a more narrow contact of different traditions and cultures in attempting to arrive at a common understanding of sustainable development. Different concepts of nature (see chapter 2), different views of the relationship between the individual and society, different religious and cultural traditions, different conceptions of justice enter into the sustainability conflicts. Solutions in this respect will require identifying the explicit and implicit contradictions and divergencies between different cultures in the five fields of conflict mentioned above and dealing with them in a constructive way.

It therefore becomes apparent that conflicts are an inseparable constituent of discussions on sustainability, of the way to make it more concrete, and of its societal implementation. It would be "harmonistic", to ignore this fact. The reproach of harmonism rightly draws our attention to the fact that disclosure of the lines of conflict is necessary in order to be able to talk "honestly" about sustainability and in order to avoid the above mentioned dangers of instrumentalization. But in order to refute this contention successfully, it would be necessary to offer advice and strategies, how the various types of conflict can be settled constructively.

8.2 Approaches to Conflict Resolution

In the following, the conceptual question, on which cognitive and conceptual basis these conflicts can and should be settled in the interest of sustainability stands in the foreground. In the form of an exaggerated confrontation (cf. on this subject in general [8]), we distinguish between a
- naturalistic conflict management in which, with scientific methods, optimal and, in a certain sense, "objective" paths to sustainability are determined, and a
- culturalistic conflict management, in which, besides scientific knowledge, societal discourses and normative reflection would play an important role.

The naturalistic concept is based on the assumption – to put it simply – that sustainability means lasting stability in the relationship between society and the environment [9]. It would then be science's responsibility to determine the carrying capacity and the critical loads of natural systems [10]. Because tolerance limits can hardly be defined empirically without exceeding them, and because this sort of empirical "test" rules itself out (because of its negative and possibly irreversible consequences in the sense of global change), a key role is ascribed to the integrative modelling and simulation of interactions between humankind and the environment [11, 12]. In this manner – one hopes –, "objective" standards for sustainability could be formulated, which would make the societal conflicts – at least in the questions treated by scientific methods – unnecessary. The conflicts, with their subjective and ideological aspects, would be decided virtually "objectively".

One has also tried to transfer the concept of "carrying capacity", which has been adopted from the ecological debate, into the societal sphere: "… the insight has been gaining ground that, in the areas of the economic and social systems as well, there are limits of load capacity, which – in the case of overtaxing – can lead to similar consequences (from loss of productivity to the collapse of the system concerned). " ([13], p. 17; critique of this standpoint in [3], chap. 4.1.2). The latitude left for humanity would, according to this conception, in principle have to be defined by the "objective" load-carrying capacity of the ecological, economic, and social systems. The determination of these limits would be science's responsibility, which could therefore decide the resulting conflicts objectively. Making sustainable development work would be a task for optimising the future path of more or less calculable natural and social systems. For the role of societal and therefore "non-objectivizable" conflict management, there would remain only the task of setting the safety margins in a so-called "guard-rail" concept.

The question is to what extent the expectations set in the naturalistic approach – to decide societal conflicts scientifically – are justified. The following problems present themselves:

- Load limits and carrying capacities can, as a rule, not be determined solely by the natural sciences (for the case of ecological problems). The problems are often of a character other than the eutrophication of a body of water by phosphates (in which case there actually is a clear limit of carrying capacity), but are rather a question of a more or less moderate increase of the risk of biohazard by certain anthropogenically-influenced input without a sharply-defined limit of load factor. For the limits of carrying capacity of social or economic systems, this holds true to a much greater extent.

- The intergenerational aspect of sustainability confronts the present generation with questions of long-term responsibility, and therefore with the question of an equitable distribution of the use of the natural and social resources through time. Questions of distributive justice can't be decided by reference to results of earth systems analysis.

- Questions of the just distribution of chances for making use of the various types of capital, especially natural resources, can't be decided naturalistically, because they involve ethical problems and concern internal questions of societal organization on the global level.

- The incompleteness and the provisional nature of (scientific) knowledge lead to the fact that societal actions with regard to sustainability always include risk. The resulting conflicts over risk acceptance can't be decided naturalistically, but require societal discourses.

It therefore turns out that the naturalistic attempt at conflict management by giving an "objectively" best solution according to sustainability aspects encounters limits at several points. The conclusion is, therefore, that exactly the central conflicts of sustainability – questions of the just inter- and intragenerational distribution of chances for utilizing natural resources, questions of priorities in conflicts inherent in sustainability, as well as questions of dealing with the inevitable problems of dealing with risk – can't be answered by the naturalistic approach. Knowledge and values can't be kept distinctly separate [14], but rather, societal values pervade even the results provided by a "sustainability science". Lines of conflict have their effects on these results – for this reason, scientific findings can't even logically be used to solve the societal conflicts.

It therefore doesn't seem surprising that, in discussions on and around the subject of "sustainability science", the level of political and normative conflicts is barely even mentioned. In the original manifest on "sustainability science" [10], the word "conflict" doesn't even occur. If the purpose is seen as providing systemic knowledge so that "social actors with different agendas can act in concert under conditions of uncertainty and limited information" [10], then the plane of legitimate political negotiation and of ethical reflection is ignored. The background assumption is presumably that the "systemic knowledge" as a result of scientific research on the relationship between the environment and society (e. g., in the form of integrative modelling within earth systems analysis [9]) decides in a naturalistic manner which actors in cases of conflict are in the right. This orientation has been tempered in further development of the concept; nonetheless, this question still has to be discussed further [15].

This brings up the question of the culturalistic approaches to conflict management. Here, we can distinguish at least three different schools of thought:
a) the political-decisionistic approach leaves the decision in societal conflicts to the political system;
b) the discursive-participative approach relies on organizing broad societal communication;
c) the ethical/justice-theoretical approach offers conflict management on the basis of universal ethical principles.

These approaches have their respective strengths and weaknesses, which can't be discussed here in detail ([16] for the case of engineering ethics). We only want to call to mind the facts that the decisionistic approach contradicts the polycentric self- understanding of modern societies, that the ethical/justice-theoretical approach is fraught with problems of implementation, and that, with regard to the discursive-participative approach, the question poses itself, why societal dialogues, in which decisions about acceptability are made, and in which power and interest constellations build the foundations, should work to the advantage of sustainability, if the actors' egoism and preoccupation with the present play a decisive role ([17], p. 32 ff.) There is no royal road to the solution of these problems, but their solution would require a combination of different approaches from one case to another.

In particular, it is insufficient, even in the culturalistic perspective, merely to refer to the societal dialogue on sustainability. Leaving this dialogue to itself would mean refraining from doing everything in one's power to make use of available potentials for rationality in acquiring knowledge and in seeking orientation. The societal dialogue and its organization, down to the establishment of formal, legitimized decision-making procedures, determine, in the final analysis, in which manner the lines of conflict within the concept of sustainability can be broken up. What is decisive, however, is that these processes are not run "blindly", but are "informed": informed by the results of interlinked models, by knowledge about the systems involved, by the knowledge about the impacts of human activities; informed, too, about the "culturalistic" components of the conflicts, namely by ethics, by the theory of justice, and by the social sciences.

The question is therefore not that of an objective-naturalistic transition to sustainability [10], but of the scientific accompaniment of a societal process of gaining awareness, opinion formation, conflict management, and decision-making, in which sustainability is, normatively, first constituted – with observance of the ethical dimension of responsibility for the future.

In order to be able to propose some more concrete ideas to enter this way of proceeding, it is necessary to explain the basic understanding of sustainable development in more detail.

8.3   The Integrative Concept of Sustainable Development as a Framework for Cultural Conflict Resolution

The sustainability concept interweaves two lines of discussion which have been rife since the 1970's. These are, first, the discussion of the threat to the natural environment through human intervention (climate change, the ozone hole, etc.) and about the corresponding efforts for more environmental compatibility. This direction is the focus of the sustainability debate in the industrialized nations. Second, and this is, above all, that which in the developing countries is understood under "sustainability", there is the political question of a just world order between developed and developing countries [1]. Sustainability can not be reduced to the environmental dimension alone but has societal, economic, and political aspects [1, 3]. In order to meet this challenge, and to uphold the concept of sustainability's meaningfulness in spite of these dilations, an integrative concept of sustainable development has been elaborated within the framework of the HGF joint project "Global Sustainable Development – Perspectives for Germany"[18]. This concept will be described in short below as an important example for operationalization of the normative aspect of sustainability (s. [3] for greater detail).

8.3.1   The Integrative Concept – An Overview

The well-known definition formulated by the Brundtland Commission [1], according to which development is sustainable when it "meets the needs of the present generation without compromising the ability of future generations to meet their own needs", the Commission's explanatory texts, and other central documents of the sustainability discussion, such as the Rio Declaration, enable us to define the following constitutive elements of sustainability ([3], Chap. 4):

- Justice: Sustainability and justice are inseparably interwoven. In particular, inter- and intragenerational aspects of equity are both equally constitutive for sustainability.
- The Planetary Scope: The global perspective and problems of global scope are the points of departure for defining criteria for sustainability.
- Anthropocentrism: Anthropocentric premises have from the beginning been inherent in the sustainability discussion, because sustainability is a question of the human use of resources.

On this basis, the following super-ordinate goals of sustainability can be formulated, which take a first step in the direction of concretization [3]:

- securing mankind's existence,
- upholding society's productive potential,
- keeping options for action and development open.

These goals are more clearly concretized in the next step by sustainability rules, which apply to various societal areas, or to certain aspects in the relationship between society and the natural environment (Table 8.1). Such material "What"-Rules present minimum requirements on content for attaining the three general goals.

Further, instrumental "How"-Rules were also formulated, which concern the method of fulfilling these minimum requirements ([3], Chapter 6). Here, it is a question of which basic institutional, political, and economic conditions have to be given in order to be able to put sustainable development into practice, that is, to promote its realization (Table 8.2).

Table 8.1 The Three Superordinate Goals and Their Respective Minimum Essential Requirements (the "What"-Rules of Sustainability)

Goals	Securing Mankind's Existence	Upholding Society's Productive Potential	Keeping Options for Development and Action Open
Rules	(1) Protection of Human Health	(1) Sustainable Use of Renewable Resources	(1) Equal Access to Education, Information, and Occupation
	(2) Securing the Satisfaction of Basic Needs	(2) Sustainable Use of Non-Renewable Resources	(2) Participation in Societal Decision-Making Processes
	(3) Autonomous Self-Support	(3) Sustainable Use of the Environment as a Sink	(3) Conservation of the Cultural Heritage and of Cultural Diversity
	(4) Just Distribution of Chances for Using Natural Resources	(4) Avoidance of Unacceptable Technical Risks	(4) Conservation of Nature's Cultural Functions
	(5) Compensation of Extreme Differences in Income and Wealth	(5) Sustainable Development of Real, Human, and Knowledge Capital	(5) Conservation of "Social Resources"

Table 8.2 The instrumental rules of sustainable development (the "How-Rules")

Short title	Full description
Internalization of external social and environmental costs	Prices have to reflect the external environmental and social costs arising through the economic process.
Adequate discounting	Neither future nor present generations should be discriminated through discounting.
Debt	In order to avoid restricting the state's future freedom of action. its current consumption expenditures have to be financed, as a matter of principle, by current income.
Fair international economic relations	International economic relations have to be so organized that fair participation in the economic process is possible for economic actors of all nations.
Encouragement of international cooperation	The various actors (government, private enterprises, non-governmental organizations) have to work together in the spirit of global partnership with the aim of establishing the prerequisites for the intiation and realization of sustainable development.
Society's ability to respond	Society's ability to react to problems in the natural and human sphere has to be improved by means of the appropriate institutional innovations.
Society's reflexivity	Institutional arrangements have to be developed, which make a reflection of options of societal action possible, which extend beyond the limits of particular problem areas and individual aspects of problems.
Self-Management	Society's ability to lead itself in the direction of futurable development has to be improved.
Self-Organization	The potentials of societal actors for self-organization have to be increased.
Balance of Power	Processes of opinion formation, negotiation, and decision-making have to be organized in a manner which distributes fairly the opportunities of the various actors to express their opinions and to take influence, and makes the procedures employed to this purpose transparent.

The ecological, economic, social, and political dimensions of sustainability are treated as equal in this system of rules. Improvement of the economic and social living conditions should – on the global level – be brought into harmony with long-term protection of the natural basis of subsistence. The rules are, on the one hand, supposed to serve as guidelines for the further operationalization of sustainability (e. g., to determine suitable indicators for monitoring observance of the rules (cf. [3], chap. 7]. On the other hand, they also have the function of normative criteria of supervision with the help of which conditions or developments can be judged with regard to sustainability. The requirement for making distinctions through the concept of sustainability, as expressed at the outset, is therefore fulfilled by this system of rules.

Conflicts of goals between rules can exist on various levels. First of all, it is not to be ruled out that – due to the concrete situation – simultaneous observance of (all of) the rules, is not possible. Undiminished population growth, for instance, could render it impossible to satisfy the basic needs of the world population without breaking the ecological rules of sustainability. Sustainability could then, in principle, no longer be ensured. Further, conflicting resource uses are also conceivable, which, for example, let the precept "to preserve landscapes of especially characteristic nature and beauty" come into conflict with the demand for securing an independent livelihood. Other conflict potentials can arise whenever it is a matter of translating the sustainability precepts inherent in the rules into concrete responsibilities of action for societal actors. In such conflicts, each rule can be valid only within the limits set by the others. There has to be an essence which may not be disregarded. For instance, the postulate of securing the subsistence level for all people, can, depending on the respective national context, be interpreted quite differently – in its essence, however, it merely stipulates that everyone must at least be able to survive.

The conflict potential included in the sustainability rules shows that even an integrative concept of sustainable development is not a harmonistic concept in the above-mentioned sense (cf. part 1). Rather it leads to the situation that many well known societal conflicts can be seen as sustainability conflicts. The integrative nature of sustainability increases the number of relevant conflicts because more aspects and issues are regarded as sustainability aspects and issues so that the number conflict-generating combinations also increase. This approach is "honest" in the way that it is able to uncover those – otherwise hidden and tacit – conflicts in defining, concretising and implementing sustainable development. This line of thought supports the central thesis of this paper, namely that conflicts in the sustainability debate are by no means to be avoided or seen as disturbing elements but rather are at the heart of any activities to make sustainability work. It becomes obvious that conflicts are inherent in every stipulation of sustainability strategies, and that rational conflict management is, therefore, of great importance.

8.3.2 Conclusions Drawn for Cultural Conflict Management

The system of sustainability rules, proposed by the integrative approach, allows deriving some requirements and conclusions for the constructive and "sustainable" management of conflicts which have been uncovered, strengthened or even created by discussions over sustainable development. These conclusions may be drawn from some "what"-rules as well as from some relevant "how"-rules of sustainability. The following "what"-rules are primarily affected:

Rule of Equity: A minimum prerequisite for sustainable development would be the guarantee of equal opportunity in access to education, information, culture, to an occupation, to public office, and to social status. Free access to these goods is seen as the basis for equal opportunity for all members of society to develop their own talents and to realize their life plans. As a basic precondition for a self-determined life, equal opportunity is, at the same time, a necessary prerequisite for meeting the demand for autonomously earning one's own living, and to be able to engage in societal debates on how to make sustainability work in a more concrete manner, as well as to engage in societal conflicts and their solution.

Rule of Participation: The second indispensable minimum requirement is the opportunity for participation in relevant decision-making processes. The basis for this rule is the conviction that a society can only be considered sustainable – in normative as well as functional respect – when it offers its members the opportunity for participation in the formation of societal volition. Its purpose is to uphold, broaden, and improve democratic forms of decision-making and conflict management, especially in view of decisions which are of critical importance for the future development and organization of society. In the concept of sustainability, participation is a means as well as an end: with regard to the individual's right to a self-determined life, participation is a goal. Proceeding on the conviction that a process of development in the direction of sustainability can only be successful if it is initiated and supported by a broad societal basis, participation is, at the same time, an instrument. The role of participation for a "sustainable" management of societal conflicts is self-evident.

Rule of the conservation of the Cultural Heritage and of Cultural Diversity: A further minimum requirement is that the historical heritage, as well as the diversity of cultural and aesthetic values is preserved. This precept includes the protection of nature above and beyond its economic function as a source of raw materials and as a sink for pollutants: nature, that is, certain elements of nature, must be protected because of its cultural importance as an object of contemplative, intellectual, religious and aesthetic experiences. Conflict management, therefore, should take into account the plurality of cultural traditions involved, as well as the experience with approaches to conflict management known from history.

Rule of conservation and development of social resources: The expectations of individuals with regard to self-actualization and autonomy don't necessarily harmonize with society's demands for integration, stability, and conformity. In the interest of sustainable development, this conflict relationship must be balanced out. A society which wants to remain lastingly viable has to provide for the integration, socialization, participation, and motivation of its members, and have the capability of appropriate reaction to changed circumstances. A minimum requirement for securing society's cohesion is seen in maintaining its "social resources". This means that tolerance, solidarity, a sense of civility and justice, as well as the capability for the peaceful resolution of conflicts must be improved – factors that are, obviously, also of high importance for the chances of solving sustainability conflicts in a constructive manner.

Furthermore, constructive conflict management is, obviously, a communicative endeavour. Conflict management can be made by argumentation or by bargaining, by different forms of negotiation and mediation, by participative approaches and by broad debate. In this respect the integrative concept of sustainable development provides some advice in which direction such communicative procedures should be developed and applied for sustainability conflicts and for institutions supporting those types of "sustainable communication patterns". As can be found in Table 8.2 the following instrumental rules are relevant:

- Society's ability to respond
- Society's reflexivity
- Self-Organization
- Balance of Power

It is of crucial importance that the impacts and consequences of approaches to make sustainability work are uncovered (reflexivity) and that an open debate is possible (balance of power). Furthermore, it has to be ensured that the results of those reflexive processes can be taken into account in the relevant decision-making processes (ability to respond). The solution of conflicts at the level of people directly concerned is preferred compared to top-down-approaches (self-organisation, subsidiarity).

In this way, it becomes clear that approaching sustainable development is a societal endeavour for which cultural resources are needed in a double manner. At first, they are needed as inputs to the debates, as standpoints and experiences in the conflicts and discussions about sustainability. At second, a new "culture" of conflict resolution is required which extends existing experience because of the extension of the scope of sustainability in time and space, compared to other issues in ongoing debates. This situation forms a formidable challenge for all societal groups.

8.4 Research Perspectives

This has considerable consequences, amongst others, for the role of the humanities and for the need for further research. Obviously, deficits of knowledge about the relationship between humankind and the environment have to be filled, systemic relationships of the human economic system have to be investigated, the foreseeable effects of societal-political interventions in even more complex cause-and-effect relationships have to be studied, modelled, and simulated, in order to make forecasts about the potential success of measures under the aspect of sustainability possible. All of this knowledge, which can often only be provided by integrative modelling, is indispensable for a policy of sustainability.

But – and the deliberations on sustainability conflicts and how to manage them point to this insight – this alone isn't enough. Many, if not the majority, of the sustainability conflicts on the various levels mentioned can't be decided naturalistically (cp. part 2). They much rather require an open societal discussion, informed by the sciences and the humanities. As soon as it is a question of conflicts based on divergent conceptions of humanity, plans for the future, and ideas of a good society, ethics as well as the social and the political sciences are called for (on the level of negotiations) to contribute to successful and peaceable conflict management.

Conflict management on the levels of the sustainability discussion mentioned in Part 2 can itself be understood as a process oriented on the imperative of sustainability. The instrumental rules of sustainability ([3], Chapter 6) show that demands for self-organization, reflectiveness, and the balance of power have consequences for the manner in which the corresponding conflicts should be settled (these rules are obviously far removed from any sort of naturalism). Sustainability's demands for equal opportunity and participation are in this respect also not inconsequential.

In sum, the following requirements for the formulation of a culturalistic concept for the management of sustainability conflicts follow out of the above discussion:

- Conflict management has to be carried out in a participative and discursive manner, in keeping with the corresponding provisions of the Rio documents. This requirement forms an inseparable component of sustainability.
- In accordance with the instrumental rules of sustainability, the appropriate instrumental and political frameworks have to be established for this purpose. Especially, this leads to the requirement – following the sustainability rule of reflexivity (Table 2) –, that societal processes of decision-making have to ensure that enough time and resources are available to the effect that careful reflection ex ante is possible and that there are opportunities to feed the results of that reflection back into the decision-making process.
- Negotiation of the conflicts has to be informed by comprehensive knowledge of the consequences (e. g., of the foreseeable effects of unsustainable developments, of the implementation of measures, or of societal transformation, and be based on knowledge. This implies an important role for the sciences and humanities.
- In normative respects, they also have to be oriented on ethical advice (e. g., with regard to responsibility for the future, justice, and distributive problems): the cooperation between philosophical ethics, the social sciences dealing empirically with conflict management and extra-scientific actors in the field becomes more important.
- Input from and engagement by the various societal groups is indispensable. Especially the world religions are obliged to bring their experience concerning humankind and history into such conflict-solving processes.

These requirements indicate the dimensions of the challenges – challenges to societal communication and to the societal dialogue as well as to its comprehensive, interdisciplinary scientific support, which extends from research on natural systems and anthropogenic influences on them to ethics and conflict research.

8.5   Summary

The imperative of sustainability has often been criticized to the effect that too much is read into it, that it doesn't exclude anything, that it permits no differentiation, and that it generates a false sense of harmony. In this paper, the thesis is proposed that, in contrary, conflicts arise on all of the levels of making the concept of sustainable development work – not only when concrete political measures are put to debate. Several types of conflict are uncovered. Such conflicts are, regularly, rooted in plural societal values, different images of humankind and nature and different ideas about future society. The paper suggests that these conflicts could and should be used to define the more concrete societal understanding of sustainability and ways to approach it. Therefore, such conflicts are to be made transparent. The approach of settling them should be understood as an essential component of the societal constitution of the content and the interpretation of sustainability. It is stated and argued for that new cultures of conflict-solving should be established across existing cultures and traditions. This would lead to the result that, in the scientific occupation with sustainability, a new field of activity is opened for those disciplines which can contribute to this objective, such as the cultural, social and political sciences, jurisprudence, psychology, and ethics, while, at the other side, it can be shown that there are limitations to the conflict-solving capacities of natural sciences and earth system analysis.

References

[1] Brundtland Commission, World Commission on Development and Environment: Our common future, Oxford University Press, 1987.

[2] A. Knaus, and O. Renn, Den Gipfel vor Augen, Unterwegs in eine nachhaltige Zukunft. Marburg, 1998.

[3] J. Kopfmüller, V. Brandl, J. Jörissen, M. Paetau, G. Banse, R., Coenen, A. Grunwald, Nachhaltige Entwicklung integrativ betrachtet. Konstitutive Elemente, Regeln, Indikatoren, Edition Sigma, 2001.

[4] H. Daly, Wirtschaft jenseits von Wachstum. Die Volkswirtschaftslehre nachhaltiger Entwicklung, Salzburg, 1999.

[5] K. Ott, Natur und Kultur 2001, 2, pp. 55–75.

[6] I. Serageldin, and A. Steer in Making Development Sustainable: From Concepts to Action. World Bank Environmentally Sustainable Development Occasional Paper Series No. 2 (Eds. I. Serageldin, A. Steer), Washington, D. C., 1994, pp. 30–32.

[7] J. Kopfmüller in Den globalen Wandel gestalten. Forschung und Politik für einen nachhaltigen globalen Wandel (Ed. J. Kopfmüller), 2003, Edition Sigma.

[8] D. Hartmann, and P. Janich in Methodischer Kulturalismus, Zwischen Naturalismus und Postmoderne, Suhrkamp (Eds. D. Hartmann, P. Janich), 1996, pp. 9–69.

[9] H.-J. Schellnhuber, and V. Wenzel in Earth Systems Analysis, Integrating Science for Sustainability (Eds. H.-J. Schellnhuber, V. Wenzel), Springer, 1999.

[10] R. W. Kates, W. C.Clark, R. Corell, J. M. Hall, C. Jaeger, I. Lowe,, J. J.McCarthy, H. J. Schellnhuber, B. Bolin, N. M. Dickson, S. Faucheux, G. C. Gallopin, A. Grübler, B. Huntley, J. Jäger, N. S. Jodha, R. E. Kasperson, A. Mabogunje, P. Matson, H. Mooney, B. Moore, T. O'Riordan, U. Svedin, Sustainability science. Science 2002, 292, pp. 641–642.

[11] J. Rotmans in Earth Systems Analysis, Integrating Science for Sustainability (Eds. H.-J. Schellnhuber, V. Wenzel), Springer, 1999, pp. 421- 450.

[12] J. Alcamo in Integrative Modellierung für Nachhaltige Entwicklung (Eds. C. F. Gethmann, S. Lingner), Heidelberg, 2002, pp. 3–14.

[13] Enquete-Kommission des 13. Deutschen Bundestages "Schutz des Menschen und der Umwelt (1998): Abschlussbericht. Bundestagsdrucksache 13/11/200.

[14] S. Funtowitz, and J. Ravetz in Science, Politics and Morality (Ed. R. von Schomberg), Kluwer Academic Publisher, 1993.

[15]   A. Grunwald, and S. Lingner in Integrative Modellierung für Nachhaltige Entwicklung (Eds. C. F. Gethmann, S. Lingner), Heidelberg, 2002, pp. 71–106.

[16]   A. Grunwald, Science and Engineering, Ethics 2000, 6, 181–196.

[17]   K.-W. Brand, and V. Fürst in Politik der Nachhaltigkeit, Voraussetzungen, Probleme, Chancen – eine kritische Diskussion (K.-W. Brand), Edition Sigma,, 2002, pp. 15–110.

[18]   R. Coenen, and A. Grunwald in Analysen und Wege zu ihrer Bewältigung (Eds. R. Coenen, A. Grunwald), Edition Sigma, 2003.

9 Sustainability through Science-Technology-Society Education

David Devraj Kumar

Florida Atlantic University, College of Education, 2912 College Avenue, Davie, FL 33314, USA, david@fau.edu

9.1 Introduction

How to sustain the natural resources of planet earth is a major question of this millennium. According to a statement of the world's scientific academies [1], "sustainability implies meeting current human needs while preserving the environment and natural resources needed by future generations" (p. 1). The most prevalent school of thought views sustainability as a science that deals with "dynamic interactions between nature and society" by bringing together various disciplines and cultures across the world [2, p. 8059]. However, as Sarewitz [3] pointed out, factors beyond "scientific knowledge" are needed to resolve issues originating from global climate change, toxic waste, and nuclear energy. In this context the role of education in dealing with sustainability issues cannot be overlooked as "education is an essential element of all aspects of a transition to sustainability" [1, p. 2]. Sustainability education could cultivate in students interdisciplinary processes, as well as thinking and problem solving skills essential to participate in sustainable development [4]. Approaches to improving student understanding of the interactions between science, technology and society/environment must be an integral part of education. This chapter will address sustainability from a science-technology-society (STS) education perspective with a classroom example.

9.2 What is Science-Technology-Society Education?

Science-Technology-Society (STS) education as a hybrid area of scholarship deals with complex interactions – issues, concerns and problems – between science, technology and society [5]. STS education enables students to focus on societal issues originating from

Global Sustainability. Edited by P. A. Wilderer, E. D. Schroeder, H. Kopp
Copyright © 2005 WILEY-VCH Verlag GmbH & Co. KGaA, Weinheim
ISBN: 3-527-31236-6

science and technology, and provides the pedagogical latitude for students to take part in identifying the origins and formulating solutions. It "draws on a range of intellectual sources: scientists and engineers seeking more than textbook treatments; educators focused on content that matches pedagogy; and social scientists who insist that 'context' imbues the science and technology with values, politics, and consequences" (p. 2). A substantial portion of STS education is devoted to environmental topics with explicit and implicit applications to sustainability.

STS education provides a platform for analyzing science, technology and societal issues, and exploring solutions for environmental problems, leading to a variety of learning outcomes [5, 6, 7]. The learning outcomes include: problem solving and critical thinking skills (collecting, analyzing and interpreting data, inferring, testing hypotheses, decision making, evaluating); language arts skills (analyzing articles, developing curriculum materials, communicating findings, writing letters and proposals); mathematics skills (random sampling, pattern recognition, graphing and constructing scale models); and social studies skills (historical analyses, dealing with environmental and human rights issues, analyzing public policy, map reading skills) [6]. The idea of creating science education programs to further student understanding of science, technology and society interactions received much attention in the United Kingdom leading to the Social-Economic-Technological curriculum reform movement [7]. STS is a part of the National Science Education Standards [8] in the United States, and part of science curriculum in many nations including Canada, Australia, India, the Netherlands, Japan, Ghana, and Papua New Guinea.

9.2.1 STS and Sustainability

A review of state science curriculum frameworks in the USA revealed that, for example, the science education standards "science as human endeavors" and "environmental quality" are represented in 48% of the frameworks [9]. In Australia, courses dealing with transforming raw materials to meet human needs and reshaping products to meet societal needs are examples of offerings dealing with sustainability [10]. Fuel conservation, use of renewable energy sources, energy conservation and efficiency are part of teaching science with relevance to societal needs in Netherlands [11]. The Hong Kong Curriculum Development Council [12] emphasizes exploring alternative energy, controlling air pollution, and studying the effects of detergents on local water quality as part of science education in Hong Kong. Student-oriented outcomes of STS include an increased level of understanding of the pressing issues stemming from science and technology, a heightened sense of awareness of the impact of technology, and unprecedented opportunities to make informed local decisions governing science and technology in society [13].

Kahn [14] observed that local cultural perspectives influence science practices, which is particularly obvious in developing nations. In dealing with vulnerability assessments of complex human-environment systems, Turner et al. [15] address the importance of "local stakeholders voice" in sustainability issues. In developing nations, mere adoption of western science practices may not be to the advantage of local cultural issues in sustainability. STS approaches by contrast are culturally sensitive. For example, depletion of clean drinking water in agrarian communities with high levels of ground water pollution due to uninformed agricultural practices (e.g., overuse of pesticides) is an ideal STS topic for addressing culture specific sustainability development issues.

However, such issues are not limited to developing nations. As Wright [6] stated, in the Great Plains of the USA "tremendous amounts of fossil fuel energy are expended in ploughing, cultivating, and spraying chemicals on the ecosystem to destroy weeds and insect pests. The water, the soil and even the air are negatively affected. The midwest states have always advertised as having crystal clear air quality. The situation has now changed and the states are now facing many environmental and societal problems" (p. 70). Issues such as "frenzied" consumption of natural resources, greenhouse gas production, and chemical waste accumulation, are damaging more to future than present generations [16]. For example, the future of Everglades National Park is being threatened by a frenzied consumption of land for housing (due to a population explosion in South Florida), by water pollution, and by tourism.

The restoration of the Everglades is an excellent context for STS education and for preparing students to participate in sustainable developments. Restoration of Everglades National Park (a major portion of the Florida Everglades) is a critical sustainability issue where science, technology and society interact, and it is an excellent context for education. The Everglades is about 1.29 million acres of swamp, the largest legislatively protected wilderness east of the Rocky Mountains in the United States, the largest mangrove (0.23 million acres) in the western hemisphere, the largest saw grass prairie in North America, and a rich ethnographic resource of over 2000 years of human history [17]. Restoring the Everglades is a challenge, and it is critical to sustainability in the South Florida region. The Everglades is a reservoir of information and is quite ideal for developing STS educational activities in sustainability.

9.3 Sample Educational Topics Involving the Everglades

The National Park Service [17] has recommended a variety of instructional activities such as water resources, cycles in nature, finding shapes in nature, habitats in the Everglades, classification schemes, Everglades plants and animals, geography, social studies, history, endangered species, Pineland, and natural and cultural resources to promote Everglades restoration. (The activities also cover a range of vocabulary words suitable for language instruction.) A sample of topics ingrained in these instructional activities is: Water Quality (e.g., Salt-water intrusion, Macroinvertebrates – determining water quality in terms of dissolved oxygen and the presence of various invertebrates is an excellent STS activity which will be discussed later in this chapter); Endangered Species (e.g., Florida panther, American crocodile, Green turtle, Snail kite, West Indian manatee); Animal Profiles (e.g., Alligator, Bobcat, American white pelican, Short-tailed hawk, White-crowned pigeon); Non-Native Species (e.g., Australian pine and Brazilian pepper, and their effect on native mangroves and cypress); Species loss (e.g., A decline in the number of wading birds nesting in colonies due to alterations in water flow pattern caused by shellfish industry); and Cycles in nature (e.g., Water cycle, Migration of birds).

9.3.1 A Sample Instructional Activity

An STS approach to water quality is evident in The River of Life project [18] discussed below. Sherwood [18] provided the following description of The River of Life [19], an interactive multimedia simulation developed at Peabody College at Vanderbilt University as part of the NSF funded project; "The Scientists in Action Series: A Generative Approach to Authentic Scientific Inquiry" (ESI-9350510; Robert D. Sherwood, PI).

The River of Life contains three scientifically based challenges: Macroinvertebrates and Water Quality Index, Dissolved Oxygen and Water Pollution, and the Source of the Pollution and How to Clean Up a River. In addition, factors such as pH and temperature are part of the data-based simulations incorporated into The River of Life software. It also involves the "Legacy League" characters, a group of students whose mission is to help their peers solve real-world problems. The instructional activities follow six phases.

The first phase, the "Challenge Quest," is a video presentation of a challenge built around a "teachable agent" named Billy Bashinal. A discussion on water quality data collected for a class project results in concerns by his lab partner Suzie about his "just do enough to get by" attitude towards the project. The "Legacy League" senses Billy's attitude and convinces him to do more research on water quality.

In the second phase, "Generate Ideas," students are invited to tackle some questions the Legacy League ask Billy, and are given an opportunity to generate their own questions and ideas. The questions demand measures other than macroinvertebrates one could use to check one's conclusions, locate the source of the river pollution, and create steps to clean up the pollution.

The third phase, "Multiple Perspectives," provides an opportunity to listen to the League's perspectives on some of the questions. For example, "Keep in mind that fish are not macroinvertebrates and they are mobile. So, even if one is captured in your sampling net it is not counted in the water quality index," and "It takes awhile for changes in water quality to appear, make sure to look at data over several months" (p. 154), and "I, like most people, used to think a river was polluted if it had a lot of trash in it. Scientists don't see it that way. A river can look beautiful but still be polluted. It is important to understand what pollution means" [19].

The fourth phase, "Research Resources," involves background information, simulations and links to related web sites (e.g., the Isaak Walton League of America). The sampling simulations (e.g., benthic macroinvertebrates) include the Isaak Walton method of analyzing the number of macroinvertebrates and calculating water quality index, water quality measures, and gas molecules in water and air. The simulations also include data based activities such as graphing data, analyzing data and making predictions and arriving at conclusions based on data involving pH, dissolved oxygen, temperature, etc. {e.g., "Which site (A, B, or C) has the lowest WQI and which has the highest for the samples you just took?" 'Take a look at the graphs from previous years for sites A and B. What is the general trend for site A? Site B?" [19].}

The fifth phase, "Test your Mettle," involves opportunities for students to help Billy by testing themselves about their knowledge of water quality. The activities include selecting the best tool for "sampling macroinvertebrates" and "testing dissolved oxygen," "analyzing the data from his river monitoring project for his report," "making graphs for his report using data provided," and determining "where his river is being polluted," etc. [19].

In the sixth and the last phase, "Go Public," Billy's interactions with the Legacy Team, his responses to the challenges, and his analysis of data are taken into account by the classroom teacher to evaluate Billy's progress on the water quality project.

Sherwood [18] in a field test of The River of Life among fifth grade students noticed statistically significant gains in student understanding of the use of macroinvertebrates as an indirect measure of water pollution. Item by item analyses revealed this gain was due to their understanding of the relationship between the number and variety of macroinvertebrates, dissolved oxygen, and water quality.

9.4   Issues

Most of the issues about the educational aspect of sustainability originate from corporate special interests, misleading curriculum materials, and inadequate teacher training. Beder [20] systematically showed some of the adverse influences of industries on the treatment of sustainability topics in classrooms. For example, a coal association sponsored curriculum "claim[s], in the module on Coal and our Environment, that to keep coal from harming our land, air and water; coal is cleaned before it's burned. Scrubbers take out most of the harmful gases" {Lapp [21] cited in Beder [20, p. 38]}. A promotional literature by a for-profit curriculum agency (on behalf of industries) made the following sales pitch: "Kids spend 40 percent of each day in the classroom where traditional advertising can't reach them. IMAGINE millions of students discussing your product in class. Imagine their teachers presenting your organization's point of view" {cited in Beder [20, p. 37]}. According to Molnar [22] the "corporate influence on what children learn does not end with ads for products and services. American students are introduced to environmental issues as they use materials supplied by corporations who pollute the soil, air, and water" {cited in Beder [20, p. 38]}. Unfortunately, due to a shortage of science resources, increasingly teachers fall prey to such corporate sponsored curriculum materials, often available free of cost.

On the other hand, professional educator-made environmental education materials are not without blemish. An analysis of environmental education curriculum materials conducted by the Independent Commission on Environmental Literacy, ICEE [23] revealed several shortcomings. The topics reviewed by the commission included acid rain, energy, recycling, global warming, ecology, population and hunger, and forests. Some of its findings follow.

Often textbooks treat acid rain superficially. For example, "a high school textbook, Environmental Science, published in 1996 by Holt, Rinehart, and Winston, contains very little natural science in its acid rain section, but focuses on the question of acid rain impact in some detail" [23, p. 13]. The treatment of recycling as a "panacea for waste" is questionable [23]. The economic impact (e.g., glass recycling can be more expensive than energy required to produce virgin glass), and toxic impact (e.g., de-inking of news paper requires toxic chemicals) of environmental sustainability efforts are omitted. Oversimplification of the link between famine and overpopulation without making any reference to the role of the governments involved is another common misleading information found in the textbooks [23]. As deBettencourt [24] stated in a discussion of the Commission's report, the "coverage of energy and natural resources typifies another common flaw in treatment of environmental issues: Technology is presented simplistically as either the source of all our problems or the cure to all our ills. ... Renewable energy sources, by contrast, are offered as the answer if only we have sufficient political will" (p. 149). Incorporating environmental [sustainability] issues in instruction must be an academic, substantive and rigorous part of science education. Unfortunately, political and advocacy approaches to sustainability through education fail to inform students of the unintended consequences such as loss of jobs, loss of individual freedom and loss of property rights [25]. Students must be presented with both the positive and the negative aspects of the issues involved. Finally, teachers need considerable preparation in content areas such as science, mathematics and economics to be able to teach environmental topics meaningfully [23]. {In an analysis of STS implementation data across the United States, Kumar and Altschuld [25] noticed a lack of adequate teacher training as one of the impediments to successful implementation of STS approaches to teaching and learning topics related to the environment. Also, a lack of quality STS curriculum materials, preferably on the Internet is another impediment [26].}

9.5   Summary and Implications

One of the best means of addressing sustainability in classrooms is through STS education. Students need to see the connection between classroom science and the real world. In an era of test-driven curriculum aimed at increasing student scores on standardized tests, there is often very little opportunity for STS and the higher order learning skills it promotes. Furthermore, undue influences by special interest groups, and a lack of adequate curricula and teacher training are impeding the proper implementation of sustainability in education. At times it seems that corporate special interests on one side and political action groups on the other side are holding sustainability education hostage. Nevertheless, content-based rigorous STS approaches to sustainability (like The River of Life project discussed earlier in this chapter) are available if one is willing to search. It is up to scientists and educators to come together to explore ways of incorporating STS approaches to make sustainability an integral part of science teaching and leaning.

References

[1]   InterAcademy Panel on International Issues, Transition to sustainability in the 21st century: The contribution of science & technology. A statement of the world's scientific academies, can be found under
http://www4.nationalacademies.org/iap/iaphome.nsf/weblinks/SAIN.

[2]   W. C. Clark, and N. M. Dickson, Sustainability science: The emerging research program. Proceedings of the National Academy of Sciences, 100(14), 2003, pp. 8059–8061.

[3]   D. R. Sarewitz, Frontiers of illusion: Science, technology, and the politics of progress. Philadelphia: Temple University Press, 1996.

[4]   Forum on Science and Technology for Sustainability (no date). Education & Training, BLK program "21", can be found under http://sustsci.harvard.edu/edu/blk21.htm.

[5]   D. D. Kumar, and D. E. Chubin in Science, technology, & society: A sourcebook on research and practice (Eds. D. D. Kumar, D. E. Chubin). New York: Kluwer Academic/Plenum Publishers, 2000.

[6]   E. L. Wright in Science education in the rural United States, Implications for the twenty-first century, Yearbook of the Association for the Education of Teachers in Science (Ed. P. B. Otto). Columbus, OH: ERIC Clearinghouse for Science, Mathematics, and Environmental Education, 1995.

[7]   J. Solomon, Science in a social context, Great Britain: Basil Blackwell and the Association for Science Education, 1983.

[8]   National Research Council, National science education standards, Washington, DC: National Academy Press, 1996.

[9]   D. D. Kumar, and D. F. Berlin, Journal of Science Education and Technology 1998, 7(2), 191–197.

[10]   C. King, Australian Science Teachers Journal 1990, 6(3), 39–45.

[11]   P. L. Lijnse,, K. Kortland, H. M. C. Eijkelhof, D.Van Genderen, H. P. Hooymayers, Science Education 1990, 74(1), 95–103.

[12]   Hong Hong Kong Curriculum Development Council, Syllabus for chemistry, Hong Kong: Author, 1991.

[13]   R. Roy in Science, technology, & society: A sourcebook on research and practice (Eds. D. D. Kumar,, D. E. Chubin), New York: Kluwer Academic/Plenum Publishers, 2000.

[14]   M. Kahn, Studies in Science Education 1990, 18, 127–136.

[15]   B. L. Turner II, P. A. Matson, J. J. McCarthy, R. W. Corell, L.Christensen, N. Eckley, G. K. Hovelsrud-Broda, J. X. Kasperson, R. E. Kasperson, A. Luers, M. L. Martello, S. Mathiesen, R. Naylor, C. Polsky, A. Pulsipher, A. Schiller, H. Selin, N. Tyler, Illustrating the coupled human-environment system for vulnerability analysis: Three case studies. Proceedings of the National Academy of Sciences 2003, 100(14), pp. 8080–8085.

[16]   Y. Quere in The challenges for science education for the twenty-first century, Vatican City: The Pontifical Academy of Sciences, 2002.

[17]   National Park Service (no date). Everglades national park, can be found under http://www.nps.gov/ever.htm.

[18]   R. D. Sherwood, Problem based multimedia software for middle grades science: Development issues and an initial field study, A paper presented at the annual meeting of the National Association for Research in Science Teaching, St. Louis, MO, 2001.

[19]   The River of Life, Nashville, TN: Peabody College, Vanderbilt University, 2000.

[20]   S. Beder, The corporate infiltration of science education, School Science Review 1998, 80(290), pp. 37–42.

[21]   D. Lapp, Private gain, public loss. Environmental Action, Spring, 1994, pp. 14–17.

[22]   A. Molnar, Schooled for profit, Educational Leadership 1995, 53(1), pp. 70–71.

[23]   Independent Commission on Environmental Education, Are we building environmental literacy? Washington, DC: George C. Marshall Institute, 1997.

[24]   K. B. deBettencourt, in Science, technology, & society: A sourcebook on research and practice (Eds. D. D. Kumar, D. E. Chubin), New York: Kluwer Academic/Plenum Publishers, 2000.

[25]   J. C. DeFalco in Science, technology, & society: A sourcebook on research and practice (Eds. D. D. Kumar, D. E. Chubin), New York: Kluwer Academic/Plenum Publishers, 2000.

[26]   D. D. Kumar, and J. W. Altschuld, Bulletin of Science, Technology, & Society 2000, 20(2), pp. 133–138.

[27]   D. D. Kumar, Bulletin of Science, Technology & Society 2001, 21(2), pp. 95–98.

10 Sustainability and Consumption of Raw Materials in Germany

Werner Schenkel

Franklinstr. 1, 10587 Berlin, Germany, werner.schenkel@gmx.de

10.1 Religion and Sustainability are in Great Demand

The impression of me, the author of this paper, is: the more chaotic, unclear, uncontrollable the situation on earth will be, the more heaven will be included in business. The more difficult it will be to put up with reality, the more insistent the call for moral orientation and ethic values will be. After having adorned ourselves with an individual attitude for years we discovered again religion.

Also, the German Federal Minister for the Environment followed this trend. On May 6/7, 2002 he supported the religious dialogue in Göttingen planned as an orientating dialogue on the subject environment and climate.

Representatives of great world religions, Christians, Buddhists, Muslims, Jews and Bahia met to have an inter-religious dialogue the course of which and the positions held have been excellently documented [1].

This documentation points to an event held in Salzburg from January 21 to 23, 2000 on the subject:" The role of religions in view of globalize destruction." The participants in the two events agreed on the fact that the present concern about the disdain and destruction of biodiversity and the destruction of earth's atmosphere and climate encouraged them to call for restraint and responsibility and sustainable use. Yet, who is today willing to listen to such calls? Isn't it as the Chilean economist Prof. Maxneff formulated: "the world religions had the opportunity to communicate their messages and values in a period of more than two thousand years but what was the result? And the message of economy and wealth induced the peoples in less than hundred years to renounce their values and cultures and to completely rely on economic values and mechanisms. Meanness is wicked but poverty is silly. The atheistic man created an appropriate activity for himself. To have seems to be more important than to be".

For decades ethical criteria have been again and again demanded for an appropriate handling of our natural basis of life in our society. Protecting of human dignity and ef-

Global Sustainability. Edited by P. A. Wilderer, E. D. Schroeder, H. Kopp
Copyright © 2005 WILEY-VCH Verlag GmbH & Co. KGaA, Weinheim
ISBN: 3-527-31236-6

forts to be made to reach social justice are a central demand of all world religions. But what happens really? Have the specific situation of environment and the future of the natural living conditions of man improved or worsened? I think that the situation of natural environment essentially worsened. Today environmental protection may be only successful by making integral, mutual and global efforts. When placing this demand we should just remember the marathon negotiations conducted for the Kyoto protocol and the conferences on maritime law.

The conference in Göttingen ended with the abstract sentence" People of all religions agree to make a contribution to sustainability and to question the development pattern predominant in Germany." But nowhere is written what is to replace it. Wasting of goods and resources is recognized as the central evil. But nowhere is it defined what the speakers meant by wasting or who will fix the ecologically compatible standard of consumption. Nobody mentions the scourge of poverty. The industrialized countries, on the one hand, contribute essentially to destroying the environment in their own and in the third-world countries by their life style and level of consumption, but they provide also the economic prerequisites required for taking measures for environmental protection. Nobody describes the disaster which would develop if the industrialized north would actually change its working method to follow the ecological demands. Self-limitation and sustainability is demanded, yet this development should not dare be connected with economic shrinkage. We only know growth, shrinkage is not envisaged.

10.1.1 Why do We Not Use the Knowledge of the Ancients?

The reason why I take a special interest in two aspects might be my observation of the environmental scene, on the one hand, my profession as an engineer and my position as a father, on the other hand: First: What will be the use of a comprehensive knowledge if you do not behave accordingly and what will occur if we will not be able to put our knowledge into action. Our detailed knowledge improves constantly, yet it becomes ever more difficult to recognize the connections.

Thus, I have been dealing with cultures which we think were capable of a sustainable management. Such explanations I detected in the publications by King [3] and Hüttermann [4].

Let me deal in greater detail with the explanations given by Hüttermann. The specific regulations and practices for their sustainable management ancient Jews developed are astonishing. They involved instructions for preserving soil fertility, eating regulations, forest use, water use, hygiene and property of soil. Unfortunately these rules apply to an agrarian society and not to a modern industrial society. Hüttermann points out the power that emanated from this people knowing and following these rules enabled it to survive for a long time in the competition with other peoples.

Knowing the fragility of nature requires consideration, care, and not exercising of power and will to control it. How would it be possible to translate this still valid finding into specific activities in a society no longer moulded by a religion? In the Jewish society time was not money but time was the prerequisite for sustainability. Time was required for developing humus, producing compost, regenerating soil, assessing the effects of present activities on future generations.

It is astonishing that all these practices were not forced by public regulations but that the recommendations of the rabbi and the social control of the community were sufficient. This was the case even where massive conflicts arose between the individual wishes for survival and well-being and the long-term ecological considerations on sustainability (for instance the prohibition to keep small animals or goats after the state system had been destroyed by the Romans).

Today such a practice seems to be unimaginable when exceeding the maximum velocity is to be regarded as a trivial offence or utilizing opportunities to avoid taxes as clever and not as reprehensible. Whether the incapability of a society to adapt itself to ecological necessities will result in its failure to survive or in developing forces allowing to take counter-measures may be gathered from historical examples, yet nobody knows whether this will apply also to future societies.

10.1.2 The Present Economic Trends and Ecological Effects

It seems to me that the essential difference between the present society in industrialized countries and the advanced civilizations of agrarian societies is the economic trend developed today.

After the Communist system attempting to establish state capitalism broke down the private striving for profit, and the orientation towards a liberalized global market is the preferred trend. This economic system has proven to be extremely effective, enforceable and adaptable. Now the non-social market forces which destroy nature should be put in their place [9].

Though the capitalist economic system seems to have damaging effects, its social character with the respective crash barriers for environmental protection has also shown its superiority to other systems. Obviously countries organized in a capitalist way produce also the money required for taking environmental protection measures and do not only destroy environment as the former communist countries have done. The values of the participants seem to be decisive.

This statement agrees with the results of a study [7] presented by Italian economists in 2002. According to results of the World Value Survey, the basic convictions of the actors in a free market economy such as competition, personal responsibility, support of private property, trust in other people, abstinence from corruption and honesty in tax matters is more developed in devout Christians than in atheists.

The observation that capitalist action only provides the funds for protecting natural environment might correspond with the findings of Max Weber relating to the economic activities of protestants. By declaring that fulfilling worldly obligations is the highest form of moralist action Martin Luther made possible the development of modern capitalism.

10.1.3 The Prices Do Not Say Ecological Truth

During the first OPEC crisis in 1973 the result of raising the prices of raw materials without considering the economic process connected with them was demonstrated [21]. The OPEC states would never like this reaction repeated, realizing that the high oil price had detrimental effects for them. By their attitude they only made the search to make previously completely uneconomic oil sources economic, that means they only created competitors. The cross-linkage between the economic consequences of an increase in the oil price was wrongly assessed.

On the other hand, the prices of metals, phosphorus and rare earths have constantly dropped since World War II, that is, strictly speaking, the prices did not provide a signal to cope with the deficiency situation predicted by Forrester and Meadows in their report "The limits of growth". Nevertheless the utilization and re-use of these materials was given an unbelievable impetus not least by waste management [11, 12]. Also, an additional raising of the prices through taxes, levies or storage costs was not necessary here.

For another resource, fresh water, not the price is decisive for saving but obviously the inclusion of fresh water management in the social network of many countries.

We observe that the natural limits of non-renewable resources were put off far to the 21st century with the question of what will happen to unused raw materials remaining unanswered.

According to this more general observation dealing with respective developments which are occurring in Germany seems to be quite interesting [13, 14, 15].

10.1.4 Existence of the Possibility to Decouple Economic Growth from the Ecologic Destruction?

From the documents of the German Federal Bureau of Statistics relating to environmental economy accounting 2002 [15] there follows that since 1991:
- air pollutants declined to 58.1 per-cent
- greenhouse gases to 84.8 per cent
- the increase in CO_2 emissions by 90.1 per-cent
- the extraction and import of raw materials to 96.4 per-cent
- Whereas the land used for settlement and transport went up by 109 per-cent and the real adjusted gross domestic product increased by 115.7 per-cent.

Accordingly Germany would have succeeded in decoupling economic growth from the consumption of nature, with the exception of land use, an interesting development insignificantly reducing environmental pollution if the absolute consumption will remain on the present level.

It is important that a number of authors proceed on the fact that the availability of resources is less a decisive limitation than the consumption of energy for the products of metabolism. That is why it is required to reduce the average consumption of resources in any case to one tenth of the current value as a precautionary measure in the next fifty to hundred years [16].

In this publication the attempt is made to calculate the total consumption of materials of the economy of the FRG for 1989 that is material and ecological rucksack when extracting it without the consumption of water and air. Accordingly a total consumption of material by German economy of at least seventy-two tons per person and year was calculated, a consumption which is certainly not sustainable and would by regarded as completely inappropriate in many countries. Obviously the change of an industrialized society to a service society is connected with a reduction of the intensity of raw materials consumption but their absolute consumption remains very high and the environmental impacts connected with their extraction and processing will not decline either.

The determination of ecological rucksacks forming part of the proper consumption of raw materials and energy was the subject of a comprehensive research project of the BGR (Bundesanstalt für Geowissenschaften und Rohstoffe = Federal Institute of Geosciences and Raw Materials) and the UBA (Federal Environmental Agency) [17].

This makes clear that returning residues, for instance liquid manure to the suppliers of maize and soy, and compost to fruit producers, will not work in a globalize economy. The importing countries will necessarily become sinks for pollutants that contaminate their environment with pollutants and nutrients. Furthermore, our demand for flowers, fruit and vegetable creates ecological problems in the producer countries: for instance creating an additional demand for water, packaging, transport capacities to cover the demand for mobility and pesticides. All of these are environmental impacts that are no perceived in a importing country. We could speak of a virtual environmental impact which would have to be included in the figures of national consumption. The whole demand for raw materials and goods including the ecological rucksacks should be considered.

The environmental costs thus arising are nearly not allocated to the products demanded. Though being aware of this we may behave completely differently.

These connections have only gradually become clear in the past twenty years. Notably the fears of the seventies that our economic system, nearly 100 per-cent dependent on imports, would be endangered by the lack of raw materials have proven to be unsubstantiated in retrospect [20, 21]. As long as we have been a solvent peaceful partner each supplier was pleased to sell us the raw materials required.

In general, these raw materials" do not run out" but change their concentration, position and temporal availability. When cost of production becomes uneconomic, substitutes will be found. Göller and Weinberg are of the opinion that in spite of raw materials being exhausted mankind will be able to survive in a respectable way. Yet, the prerequisite will be that sufficient energy should be available to allow transformations and substitution [20].

I could even quote Prof. Einstein who is to have said "The only resource which will not be reduced by using it but will grow is imagination."

But I state that the fossil energy carrier's, water, and phosphorus are not substitutable and will be exhausted dramatically fast. I do not know whether imagination will then help us over the problems arising but, of course, I hope so. That is why I think arbitrary substitutability to be a mistake.

The German Bundestag has been intensively dealing with substance-related cycles in the industrialized society in a commission of inquiry of the twelfth and thirteenth legislative periods, thus laying the foundations for operate sustainability.

10.1.5 Sustainability as a Survival Strategy

In the twelfth legislative period the German Bundestag set up a commission of inquiry entitled "Protection of man and environment – criteria of assessment and prospects of ecologically compatible substance-related cycles in the industrialized society" establishing the following rules of management of resources[10]:
1. The rate of extraction shall not exceed the rate of regeneration.
2. Use non-regenerative resources as economically as possible and only to the extent as a physically and functionally equivalent substitute will be provided in the form of renewable resources or a higher productivity of the non-renewable resources.
3. Substance inputs into environment shall orientate themselves by the maximum impacts on environmental media considering all functions.
4. The time of anthropogenic inputs into environment and disturbances of it shall be in a balanced ratio to the time of the reactivity of environment-relevant natural processes.

In the thirteenth legislative period this commission of inquiry modified its subject of work" Protection of man and environmental targets and basic conditions of a sustainable prospective development". In this report from the German Bundestag the sustainability concept was formulated [10].

These principles result in decoupling economic development from the consumption of non-renewable resources including proposals relating to the reduction of the absolute quantity. I quote four sources:
1. The report on a promising Germany by the Wuppertal Institute [11].
2. The two reports by the Federal Environmental Agency" Sustainable Germany – ways to a sustainable and ecologically just development "[12] and "Sustainable Germany – organizing future permanently ecologically just [13]".
3. The strategy for a sustainable development for Germany [14]. Positions of the Federal government for the UN conference in Johannesburg 2002 (Rio+10).

In all four papers it is notable that they do not make comments on the central subjects: private property and return on capital. In my view these papers lag behind the ancient Jewish solution mentioned above. Yet, they demonstrate excellently the alternative scopes in the developed countries.

Comments on the consequences of, the finite nature of resources, for the economy are rather disputed. After the predictions by Meadows and Forrester in the first report from the Club of Rome on the limited availability of resources did not come true the experts agreed on the fact that fossil fuels, water and phosphorus are to be considered as especially worth being protected whereas other raw materials are to be considered to be substitutable by intelligence and imagination so that their consumption is regarded to be less dangerous for sustainability. Because what shall those who will inherit the resources do other than using and consuming them. Thus, a shortage of these resources would give rise to innovations that must necessarily be dealt with in our society and in the world religions. Progress brought: wealth, security, comfort, health, mobility, education, information. In short: In our view, it made life pleasant. Who would thereupon like shrinkage instead of growth? Progress brought technological and economic growth. Yet, in assessing this technological progress we could be greatly mistaken. May we voluntarily limit progress or shall we also do what we can in future? What is the significance of the future technological progress in the life style of sustainability? As any technological progress involves social implementation, this progress always infers societal change. Only its social integration shows whether the progress will be considered to be beneficial or obstructive and by whom.

10.1.6 Worldwide Interlacing Will Make our Prosperity Safe

In Germany a remarkable change in assessing the supply of raw materials has taken place. We waged World War II, inter alias, with the argument "people without space" and wanted to appease our hunger for crude oil in the Caucasus. Thus the undersecretary of state in the Federal Ministry for Economy and Labour and many years' manager of Krupp Mommsen pointed to the fact that one of the reasons of the glorious reconstruction in Germany after 1945 was that we were able to buy our raw materials worldwide where they were available at the lowest prices unimpeded by the consideration for colonies and Commonwealth members. In the seventies research was conducted by the BGR to use and stockpile especially rare raw materials [18]. These plans were rejected and not implemented – with the exception of raw oil reserves.

Today we hold completely different views. As an economically strong country, we were able to buy worldwide raw materials at the lowest prices. But owing to pursuit of such a policy we became a big sink for pollutants and nutrients in spite of all exports. Whether this economic advantage will be also an ecological advantage has only to be proved.

Recently Ather D. Little established in a study that it would be possible to reduce reliably the costs of material and energy required at present in production and product use which total € 180 billion, by twenty-five per cent if respective incentives would be present, that means that here still essential possibilities of economy are to be found.

That is why it is undisputedly applicable that the MIPS concept of Dr. Schmidt-Bleek from the Wuppertal Institute to strive in future for saving of material by a factor of four to ten will be implementable [22].

The annual report of 2000 of the German Hard Coal Association reads as follows:" There is a strident disproportion between the present consumption of fossil energies and the respective stocks of energy carriers. Whereas there exist the biggest resources of coal the consumption of oil and gas is dominating. Some studies of resources expect bigger bottlenecks in oil and gas supply."

The dependence on imported oil will drastically increase in Western Europe and in the USA. The competition for imported oil between Western Europe, North America, and the Southeast Asian threshold countries poor in raw materials will further aggravate.

The OPEC states will increase their share in the world oil extraction from forty to fifty-four per cent before the year 2030 together with a predicted large increase in the oil and gas prices.

Covering the demand for raw materials will result in a competition so far unknown.

In addition, Germany formulated a strategy for a sustained development. However, in this paper the consumption of resources is not dealt with explicitly but included indirectly in a multitude of other fields of action. The targets of this strategy are comparatively modest. In my view they do not consider the development of the industrial society via the science to the information and fun societies. It does not deal either with the demographic modifications becoming apparent already. The population will essentially decline from 84 million inhabitants depending on the approach. The age pyramid will clearly shift and mould the life style of the citizens of this country and thus the desired goods, also.

10.2   Conclusions

- The question whether the availability of raw materials or their environmental impacts during their extraction, processing and consumption supports the change of paradigms desired for sustainability may not be answered.
- The present use of raw materials is faced with many modifications associated with the attractive life style. Its modification by a voluntary limitation (to have less and to live better) seems to be rather improbable.
- The ethical rules and practices relating to sustainability come mostly from religions developed by advanced agrarian civilizations. In spite of these rules the multitude of the civilizations has not survived. They have not considered the rules in their practices.
- For the present society ethical rules and practices have first to be formulated and a social consensus on them has to be reached. The consensus would have to be put into practice.

For about 6000 years all forms of life have changed our environment. Why should this be different in future? Who says that this culture has not gone beyond its climax? We did not live in a sustainable way but therefore we were able to fly to the moon.

- Moderation or voluntary limitation of the consumption of resources may not be generally substantiated by the finiteness of the resources. This does not predominantly refer to fossil energy, water, air, land and phosphates. They may not be substituted.
- For the *homo fossils* the short-term dissipation of available energy accumulated over millions of years plays a decisive part. Without the help of non-renewable energies our present world would not be imaginable and every change striven for is presumably connected with massive social changes which are to be implemented.
- When considering the efforts made by the ADAC throughout years to reduce the number of the persons injured and dying in road traffic and when considering the financial and staff resources required for that – which nevertheless were not able to essentially reduce the 490,000 personal damages occurring every year – you can imag-

ine the expenditure which would be required to make economy of raw materials and energy popular and to implement it as a social target. The expenditure required to remind people of ecological care and to practice it seems to be huge. It is still more attractive to park your Mercedes in front of your house than putting it on your house roof in the form of a solar plant.

- I think the sentences by Hardt and Negris:" We reply to the misery of power with the pleasure of being" seems to be arrogant and not acceptable for the majority of the population.
- A limitation to meeting necessary human wants and a moderate, careful use of natural Resources seems to me desirable but unrealistic.
- Recycling is a value with the character of redemption rather distracting from the real conditions than contributing to their improvement. Recycling does not pay an essential contribution to resources conservation. Nevertheless we should strive for establishing circulations in our economy.

Contribution to resource establishing a circulation in the present society is not sustainable, it is globally cross-linked and regionally obsolete, it is economically orientated and internationally in business. It orientates itself towards a world culture and turns the world into a village. This society may no longer be controlled by old religious values and orientations. We would like to have an up-to-date orientation for action accepted everywhere also if we eventually will not conform to it.

- In Germany a sustainable development has not yet been reached in spite of many successes achieved in environmental protection though the efficiency of resource use has been increased in many cases. The consumption of resources per produced unit of goods and services has consistently decreased; however the saving effect connected with it was improved by clearly increasing the production and use of these goods and services. In this way decoupling of economic growth from the consumption of resources was achieved but an absolute decline of the consumption of resources is still to be expected.
- Without water and air but with other ecological rucksacks the German consumption of raw materials totals seventy-two tons/person a year according to the inquiries made by the Federal Bureau of Statistics. This value is not sustainable. With the world population totalling 6 billion people and consumption as in Germany – which certainly is not applicable in this way – the yearly demand, without use of air and water, would total five hundred sixty billion tons/year, a number which is completely non-realistic and would result in most heavy damages.
- Shrinkage to reach a new quality of the quantitative growth to allow new growth by metamorphoses, for the time being, is considered to be unrealistic. Yet, it seems to me that it will be the only solution of the problem.
- According to the view of the founders of religions the world is dynamic. If mankind will not fulfil its assignment it will loose its assignment (for example also by animals – insects, bacteria, viruses).
- Now, as before, we are in search for a sustainable managed world population in the next fifty years. Man forms part of nature and are not its controller that means the technological, scientific and economic activities would have to be accordingly included in the ecological conditions. We need a change of paradigms in our thinking

and culture in the next fifty years. In this connection preserving of nature is less a concern than the preservation of the anthropogenic ecosystems. We will have to develop new ways and not to continue tried and tested ways.

References

[1]   G. Orth, Die Erde – lebensfreundlicher Ort für alle, LIT Verlag, 2002.

[2]   K. W. Weber, Smog über Attika – Umweltverhalten im Altertum. Arte Verlag, 1990.

[3]   F. H. King, Viertausend Jahre Landbau in China, Korea und Japan. Edition Siebeneicher, Volkswirtschaftlicher, Verlag München, 1984.

[4]   A. P. Hüttermann/A. H. Hüttermann, Am Anfang war die Ökologie – Naturverständnis im Alten Testament, Verlag Antje Kunstmann, München, 2002.

[5]   H. Kessler, Ökologischer Weltatlas im Dialog der Kulturen und Religionen. Wissenschaftliche Buchgesellschaft, Darmstadt 1996.

[6]   Chr. M. Schwald, Religionsgeprägte Weltkulturen in ökonomischen Theorien, Verlag V. Florentzen GmbH, München, 1999.

[7]   L. Ginso, P. Sapienza, L. Zingeles, Peoples Opium? Religion and Economic Attitudes. CEPR Discussion Paper no. 3588, 2000.

[8]   NN. Die Industriegesellschaft gestalten. Bericht der 12. Enquete Kommission des 12. Deutschen Bundestages, Economic Verlag, Bonn, 1994.

[9]   NN. Konzept Nachhaltigkeit – Leitbildung zur Umsetzung der Enquete Kommission "Schutz des Menschen und der Umwelt" des 13. Deutschen Bundestages, German Bundestag, 1998.

[10]   Federal Government/Misereor. Zukunftsfähiges Deutschland. Ein Beitrag zu einer global nachhaltigen Entwicklung. Wuppertal-Institut für Klima – Umwelt – Energie, Birkhäuser, Verlag, 1997.

[11]   Federal Environmental Agency, Wege zu einer dauerhaft umweltgerechten Entwicklung. Erich Schmidt Verlag, Berlin, 1997.

[12]   Federal Environmental Agency, Nachhaltige Entwicklung in Deutschland, Die Zukunft dauerhaft umweltgerecht gestalten, Erich Schmidt Verlag, Berlin, 2002.

[13]   Federal Ministry for the Environment, Nature Conservation and Nuclear Safety, Perspektiven für Deutschland, Unsere Strategie für eine nachhaltige Entwicklung.

[14]   Federal Bureau of Statistics, Umwelt – Umweltökonomische Gesamtrechnung, 2002, Federal Bureau of Statistics.

[15]   Stefan Bringezu, Neue Ansätze der Umweltstatistik, Wuppertal-Institut, 1995.

[16]   Chr. Kippenberger, Stoffmengenflüsse im Energiebedarf bei der Gewinnung ausgewählter mineralischer Rohstoffe. Geol. Jahrbuch, Reihe H, Heft SH 1, E. Schweizerbart'sche Verlagsbuchhandlung, Stuttgart, 1998.

[17]   U. Engelmann, Die Rohstoffabhängigkeit der Bundesrepublik. in Schriften des Vereins für Sozialpolitik, vol. 108. Verlag Duncker und Humboldt, Berlin, 1980.

[18]   B. Fritsch, Über die partielle Substitution von Energie, Ressourcen und Wissen. in Schriften des Vereins für Sozialpolitik, vol. 108. Verlag Duncker und Humboldt, Berlin, 1980.

[19]   BGR, Ausfallrisiko bei 31 Rohrstoffen. unpublished study, 1977.

[20]   C. W. Sames, Die Ängste der Siebziger Jahre waren unbegründet, Handelsblatt, 31.10.91, pp. 44.

11 Sustainability, Culture and Regional Scales: Some Remarks from Human Geography

Horst Kopp

Institute for Geography, Kochstr. 4/4, 91054 Erlangen, Germany,
hkopp@geographie.uni-erlangen.de

There is extensive agreement in academic and political discussion of the premise that sustainability includes the three dimensions ecology, economics and society. Sustainable development should involve a developmental approach that considers responsible utilization of natural resources and social acceptability but does not exclude economic growth. In other words, preserving the environment, growth, social justice, cultural identity and participation must be considered in terms of their interaction if the objective of sustainability is to be achieved [1].

- This discussion stemmed from the realization that a policy which is neo-liberal in character and oriented toward growth and modernization inherently causes environmental problems and social deformation:
- Poverty is not eliminated, instead it is frequently generated;
- Population pressure cannot be effectively reduced without consideration of normative aspects;
- Hasty and inappropriate implementation of exogenously developed technologies as well as excessive orientation toward economic growth result in inappropriate forms of utilization;
- Technological "progress" eliminates indigenous resource utilization strategies, aggravates social disparities and inhibits participative structures [2].
- It is therefore necessary to strike a "fair balance of ecological, economic and social interests" [3]. Because ensuring the capability to perform in an ecological, economic and social context involves very complex interdependencies and cannot be partially optimized without consideration of developmental processes in overall terms [4], integrative strategies must be found that ensure the interaction of all three dimensions. This is only possible if interdisciplinary or transdisciplinary cooperation based on a systemic-theoretical approach is taken [5]. Consequently it goes without saying that this contribution from the point of view of human geography can only constitute a single element of a comprehensive analysis of the problem.

Global Sustainability. Edited by P. A. Wilderer, E. D. Schroeder, H. Kopp
Copyright © 2005 WILEY-VCH Verlag GmbH & Co. KGaA, Weinheim
ISBN: 3-527-31236-6

It is helpful to first consider the three dimensions of sustainability from the human perspective in greater detail (cf. [5] for aspects pertaining to weighting of these dimensions):

- *Ecological sustainability* does not merely imply "preserving the natural potential" (biodiversity, survival of ecosystems) as a resource pool for human utilization, but rather also includes moral values in the sense of "respect for creation", because people/social groups associate esthetic and symbolic qualities with the environment (in the sense of nature) than cannot be measured economically and are therefore often not duly regarded [6].
- *Economic sustainability* comprises not only economic growth, combating poverty and raising the standard of living, but rather must also guarantee social acceptability [7].
- *Social sustainability* considers intergenerative and intragenerative aspects in the sense of maintaining the social capital. This includes such diverse aspects as life expectancy, participation in political activities, satisfaction of basic needs, equal chances of survival, and human rights.

Human geography regards itself a science that analyzes the human-environment relationship from a social science perspective in a spatial context. Man as an individual that takes action is at the focal point. He shapes space (his environment) by using it in many ways. The alternative courses of action are limited, on the one hand, by the available natural, technical and knowledge resources and on the other are guided by socially binding norms and values that are usually followed unconsciously. Both boundary conditions are subject to continual change, thus courses of action must permanently be adapted to this change.

Since the Cultural Turn was made by the social sciences, discussion of the issue of the "cultural embeddedness" of human actions has also become more intense in human geography. Previously geographers also spoke of "cultural regions/cultural realms" [8] to denote major regions of the world in which a "particular, unique combination of natural and cultural elements that shape landscapes" can be found that are the result of the "independent, intellectual, and social order and the context of the historic process" [9]. Edward Said suggested that "cultural realms" are imaginary concepts, because they were "devised" by scientists for political purposes and because they are based on a holistic concept of culture. Nevertheless such concepts, because they are so readily comprehensible and can be used for instrumental purposes, repeatedly undergo revivals, most recently in the case of Huntington [10].

The reality on earth is, however, far more complex. "Culture" is definitely a multilayered phenomenon, that cannot be categorized in sweeping, general terms. Every human being ultimately has his own "culture" in the sense of norms, values, material (existing and self-shaped) environment, social embeddedness and many other factors rooted in genetic heritage, socialization, and biography. Taken in this sense it also becomes clear that culture is a process of permanent change and that we are all the products of our social environment.

This social environment – and this brings us to the geographic view of culture – can in simple terms be assigned to various levels of scale. On the local scale we are influenced by family and friends, on the regional scale by the norms of social life that are based on a consensus of common values (which also includes the influence of religions), and finally on the global scale by universal human rights and to an increasing extent by "cultures" that are driven by consumption. It goes without saying that this subdivision is not dichotomous, but rather a highly complex interaction of all the above-mentioned factors. In our age of globalization the boundaries of the levels of scale are becoming increasingly unclear. Nevertheless it must be stated that our actions are always guided and determined by all of the above-mentioned aspects [11].

The discussion of globalization issues in particular has revealed that in spite of, or rather because of, the increase of the influence of global cultural elements the regional implications of an action becomes increasingly important, because the individual orients his routine courses of action toward what happens in his personal realm of experience. Societies thus obviously tend to permanently produce new regional identities.

At this point it becomes important to formulate the term "regional" more precisely. As the significance of national decision-making competencies declines, two "regional" forms are stepping more and more into the forefront: on the one hand a multi-national entity (Europe, e.g., in the sense of EU membership), and on the other smaller-scale entities of various sizes (e.g., Bavaria, Franconia, Nuremberg) that, depending on the respective context become the realm of identification and thus the level of action. It thus becomes clear from these examples that here, too, there is no clear dichotomy, particularly when "the regional" elements cannot necessarily be linked primarily to a region, but rather to a social entity (ethnic group, tribe, extended family, etc.).

Let us now return to sustainability. Geographers have concerned themselves chiefly (in the latter context described above) with regionally limited phenomena and in doing so aspects of sustainability always played a central role. This is due to the fact that interest always focuses on the relationship of man and his environment, usually from a process-oriented perspective. In the meantime, thousands of empirical studies have been conducted on issues pertaining to sustainable utilization in agriculture, problems related to sustainable urban development, sustainable forms or industry or tourism. For quite some time these research activities employed the paradigm of theoretical modernization: "traditional" structures (forms of industry, lifestyles, patterns of utilization) opposed "modern" structures and the two interacted in diverse ways. Here the question of how the respective traditions had become established was given little consideration, focus was instead on the impact of modern elements. Since the Cultural Turn the perspective has shifted. Today one speaks in a more neutral manner of "indigenous" and "other" elements in the context of the relationship of man and his environment (e.g., [12] and [13]).

It thus becomes evident that the development process of *all* societies entails a permanent interaction between endogenous and exogenous cultural elements, i.e. that social learning processes are continuously involved, which always inherently include an element of uncertainty [14]. It has become common practice to speak of vulnerability when the exogenous factors dominate over the social capabilities for adaptation, that is the case when uncertainty prevails. Such social capabilities for adaptation do, of course, vary greatly from one individual to another, but are always cultivated in a regional social context. Stated in other words: perception of new stimuli and the associated action taken in response do indeed occur on the individual level, but society contributes its system of values and social knowledge and thus establishes the boundary conditions and provides the means for deciding what action is to be taken [15].

When the discussion of sustainability is dominated by the slogan "think global – act local", as is the case particularly in the context of the Agenda 21 Process, the regional dimension is clearly lacking. Most social networks are based on this particular level of scale, the most important common sense values and norms on which actions are based are inherent to it, established social knowledge of the interrelationships between human activities and the natural environment, laws and rules hold true and adaptation to external challenges can best be met on the basis of endogenous capabilities.

This realization is well established as a fundamental component of participative developmental planning – at least theoretically. In practice, however, thoughts relating to modernization theory crop up time and time again on the "provider side", chiefly subconsciously. Further development involving implementation of sustainable utilization can only be successful when we accept the performance capability of indigenous resource utilization strategies, i.e., those endogenous potentials which as a rule have been adapted ecologically and, where expedient, in an economic context, and policy-makers ultimately support these actions ([16] and [17]).

This is, however, nothing more than obtaining a better understanding of and fully accepting *regional cultures* that only become accessible to us when the actors in the respective region are actively involved in the discussion as well. Above all these actors must be put in a position to develop capabilities that enable them to appropriately respond to the rapidly increasing challenges imposed by external factors in a manner that is based on *their own* cultural values and is in the interest of *their own* descendent generations in the sense of sustainability. The following topics for on-going, inherently problem-oriented research can be derived from this:

* Do any adapted, sustainable forms of resource utilization in fact still exist, and if so, where?
* Which socially-accepted values and norms are "responsible" for the man-environment relationship in various regional cultures?
* What role does environmental experience play in the characteristics of regional cultures?
* How flexibly do regional cultures respond to external factors without the occurrence of environmental destruction? What is responsible for this degree of flexibility?
* Is development possible in the context of certain regional cultures without technical progress?
* Are elements of a civil society mandatory for sustainable development?
* Which aspects of cultural embeddedness are required to ensure that the individual can deal with global challenges without neglecting the objective of sustainability?

References

[1]   H. G. Kastenholz, K.-H. Erdmann, M. Wolff (Eds.) Nachhaltige Entwicklung: Zu-kunftsperspektiven für Mensch und Umwelt. – Heidelberg, 1996.

[2]   National Research Council, Board on Sustainable Development, Policy Division (Ed.): Our Common Journey: A Transition toward Sustainability. – Washington, 1999.

[3]   M. Härdtlein et al. (Ed.): Nachhaltigkeit in der Landwirtschaft: Landwirtschaft im Spannungsfeld zwischen Ökologie, Ökonomie und Sozialwissenschaften (= Initiativen zum Umweltschutz 15). – Berlin, 2000.

[4]   A. Voß in Nachhaltigkeit in der Landwirtschaft: Landwirtschaft im Spannungsfeld zwischen Ökologie, Ökonomie und Sozialwissenschaften (= Initiativen zum Umwelt-schutz 15) (Ed. M. Härdtlein et al.), Berlin, 2000, pp. 53–68.

[5]   K. W. Brand in Nachhaltige Entwicklung und Transdisziplinarität: Besonderheiten, Probleme und Erfordernisse der Nachhaltigkeitsforschung (= Angewandte Umweltfor-schung 16) (Ed. K. W. Brand), Berlin, 2000, pp. 9–28.

[6]   O. Renn in Nachhaltigkeit in der Landwirtschaft: Landwirtschaft im Spannungsfeld zwischen Ökologie, Ökonomie und Sozialwissenschaften (= Initiativen zum Umwelt-schutz 15) (Ed. M. Härdtlein et al), Berlin, 2000, pp. 39–52.

[7]   M. Günter, Kriterien und Indikatoren als Instrumentarium nachhaltiger Entwicklung: Eine Untersuchung sozialer Nachhaltigkeit am Beispiel von Interessengruppen der Forstbewirtschaftung auf Trinidad (= Heidelberger Geographische Arbeiten 115). – Heidelberg, 2002.

[8]   A. Kolb in Hermann von Wissmann – Festschrift (= Tübinger Geographische Studien, Sonderband 1) (Ed. A. Leidlmair), Tübingen, 1962, pp. 42–49.

[9]   A. Kolb in Hermann von Wissmann – Festschrift (= Tübinger Geographische Studien, Sonderband 1) (Ed. A. Leidlmair), Tübingen, 1962, pp. 46.

[10]   S. P. Huntington, The Clash of Civilizations and the Remaking of World Order, New York, 1996.

[11]   C. Hey, and R. Schleicher-Tappeser, Nachhaltigkeit trotz Globalisierung: Handlungs-spielräume auf regionaler, nationaler und europäischer Ebene, Heidelberg, 1998.

[12]   E. Ehlers in Mountain Societies in Transition: Contributions to the Cultural Geography of the Karakorum (= Culture area Karakorum scientific studies 6) (Ed. A. Dittmann), Tübingen, 2000, pp. 37–63.

[13]   F. Ibrahim in Geographische Rundschau 55 (7/8), 2003, pp. 4–9.

[14]   S. Tröger, Sustainability in the Context of Natural and Social Environments: An actor-oriented interpretation from south-west Tanzania. – In: Erdkunde 56, 2002, pp. 170–183.

[15]   C. Ritz, and S. Thomas in Mountain Research and Development 21, 2001, pp. 104–108.

[16]   G. Kohlhepp, and C. Martin (Ed.) in Mensch-Umwelt-Beziehungen und nachhaltige Entwicklung in der Dritten Welt (= Tübinger Geographische Studien 119), Tübingen, 1998.

[17]   S. Böschen in Nachhaltige Entwicklung und Transdisziplinarität: Besonderheiten, Probleme und Erfordernisse der Nachhaltigkeitsforschung (= Angewandte Umweltfor-schung 16) (Ed. K. W. Brand), Berlin, 2000, pp. 47–66.

12   Sustainable Development in Asia: Traditional Ideas and Irreversible Processes

Dietmar Rothermund

Oberer Burggarten 2, 69221 Dossenheim, Germany, dietmar.rothermund@t-online.de

12.1   Introduction

The concept of sustainability emerged in the field of German forestry (see Chapter 1 of this book) and referred to the need of replanting trees keeping in mind their natural setting and the years required for their growth. The German administrator of mines and forests, von Carlowitz, who wrote about "sustainable use" of wood for the first time in 1713 was facing an obvious dilemma. In his time charcoal was used for smelting ore and he could foresee that the ore mines in his charge would have to be closed once the forest, which he also supervised, had been cut down. He had to make both ends meet and this made him think about sustainability. This concept was thus related to planned human interference with natural habitats based on a scientific assessment of balanced development. In this context sustainability was based on the assumption of the reversibility of processes. Replanting would reverse the process of the cutting down of trees, it would restore the original position of the forest, although with a certain time-lag. As we shall see, all traditional ideas of sustainability are based on this assumption. They do not reflect the irreversibility of processes, at best there may be an awareness of the negative results of inaction, i.e. if the forest is not replanted in time, the soil may be eroded making future replanting impossible. At present, however, all discussions of sustainable development must take into consideration that irreversible processes are at work which run their course and are to a large extent beyond human control. In this context sustainability means the adjustment to such processes and the trust in scientific progress which will compensate for inevitable losses, e.g. the depletion of fossil fuel – an irreversible process – may in due course become acceptable because of the discovery or invention of new sources of energy. Sustainable development thus means aiming at moving targets rather than restoring the status quo.

Global Sustainability. Edited by P. A. Wilderer, E. D. Schroeder, H. Kopp
Copyright © 2005 WILEY-VCH Verlag GmbH & Co. KGaA, Weinheim
ISBN: 3-527-31236-6

12.2   Population Growth as a Major Challenge to Sustainable Development

The amazing growth of population in the 20th century has burdened our planet very much because it has increased the consumption of all kinds of resources and led to the pollution of air and water etc. The main cause of this growth has been the control of diseases due to the scientific progress in medicine. It has, of course, been impossible to counterbalance this by a "terminator" program. The process must run its course. It appears to be irreversible, but in the long run there may be reverses as the German example of negative growth demonstrates. In Asia, however, which is now the home of more than half of mankind population, growth is still overwhelming. The demographic transition – the adjustment of birth rates to death rates – has only just begun in some Asian countries and as most Asian nations are literally young with about half of the population below the age of 30 years, it will take much time before the transition is completed. In Western Europe the transition was fairly rapid – from about 1900 to 1930. In India, however, the death rate has declined steadily since the Census of 1911 until it stood at less than 10 per 1000 in 2001, whereas the birth rate declined very slowly from 1911 to 1941, then increased again until 1961 and then decreased parallel to the death rate but keeping a considerable distance from it and amounting to about 23 to 1000 in 2001. Due to this extremely slow transition, the population of India which stood at to 340 mill. in 1947 increased to more than one billion in 2001. Poverty is one of the main reasons for this slow transition. For poor people children are the only guarantee against starvation in old age. They also contribute to the family income while they are still very young. Child labour (5 to 14 years of age) is a widespread phenomenon in India. Estimates range from about 15 to 44 mill. children at work [1]. Traditional expectations of child mortality make poor parents reluctant to adopt methods of family planning, even though more children survive nowadays than in earlier times [2].

Chinese population growth has been slower than that of India. But it also had attained much larger dimensions at an earlier stage. In 1953 the first modern census of China recorded 583 mill. people, at present there are more than 1.3 billion Chinese. This means that the Chinese population has only doubled while the Indian one has trebled in about 50 years. Family planning has been more rigorous in China in recent times. Initially, Mao was against it as he believed that added numbers would increase the power of the country [3]. He then unwittingly contributed to a check on population growth, because his Great Leap Forward of 1958 led to the worst famine in history, which is believed to have killed about 30 million people.

Improved varieties of food grains and the extension of irrigation have enabled Asian countries to feed their growing populations. To this extent there has been sustainable development, indeed. But at the same time the cooking of food has depleted the sources of firewood very rapidly. In India most of the rural poor cook their food on primitive stoves which utilise less than 10 per cent of the energy provided by the fuel consumed [4]. Women have to walk longer and longer distances to find any firewood at all. The forest cover of the country has dwindled to about 10 per cent as shown by satellite photographs, although not by official statistics which tend to be more optimistic.

In Southeast Asia the forest cover has been reduced by indiscriminate logging. While the tropical rainforest regenerates itself and manages to thrive even on very poor soil unfit for agriculture, the areas denuded of that forest quickly become degraded and cannot be replanted. This is an irreversible process of the worst kind. Satellite photographs show that since 1985 more than 17 mill. ha of forests have disappeared in Indonesia [5]. This kind of logging is driven by commercial activities, not by the reclamation of land needed for feeding the growing population. But population growth, of course, also increases the pressure on the environment. Pre-colonial Southeast Asia is supposed to have had a population of 33 mill. and was regarded as "underpopulated". In 1950 there were about 178 mill. and at present there are around 525 mill. Projections for 2025 estimate 718 mill [6].

The "Green Revolution" which has guaranteed the massive increase in foodgrain production also has had its undesirable effects. It requires heavy inputs of fertilisers and pesticides. Critics have referred to a "pesticide treadmill" as ever increasing doses of pesticides are required to cope with pesticide-resistant insects etc. Pesticides contaminate water and thus affect human health [7]. These irreversible processes of environmental degradation may in due course require heavy investment in water treatment etc. The problem with such processes is that they are cumulative and do not attract immediate attention. Those who plead for a detailed monitoring of the state of the environment may often be blamed for being "alarmist". Asian governments which are for the most part technocratic in their general attitude are annoyed by such alarmists and wish to silence them. The recourse to traditional ideas of sustainability does not impress such governments, which want to get ahead with their plans.

12.3   Traditional Ideas of Sustainability and the Technocratic Ambitions of Modern Governments

If we look for traditional ideas of "sustainability" we must be aware of the fact that such ideas did not arise in the context of planned interference and were not based on scientific studies of the environment. These ideas were also not influenced by an awareness of the irreversibility of processes. They evolved from the experience of generations living in their natural habitats to which they had to adjust. Such ideas were not articulated in "scientific" terms but in terms of a quasi-religious respect for the conditions of nature. Ritual served as a means of inter-generational communication. Taboos helped to prevent transgressions. Such rituals and taboos could, of course, appear to be "superstitions" to later generations that accept only rational explanations.

As a case in point we may mention the experience of the indigenous population of a remote region in Southern China who still practised swidden cultivation. Like swidden cultivators elsewhere, they burnt down only small plots of the forest and then moved on, giving the forest enough time to regenerate. In addition they demarkated certain parts of the forest as dedicated to the dead as burial grounds, others were off-limits because they were the abode of spirits. Sacred groves of this kind were also maintained by swidden cultivators in India. The practical consequence of such taboos was that there were reserves in which biodiversity was never disturbed by human interference. Under Mao's rule thousands of educated young people were sent to this region of Southern China to eradicate superstitions as well as the forest in order to convert it into rubber plantations. The results were disastrous. Biodiversity was greatly reduced, the rubber trees would not grow properly, because the climate was not suitable for them, the indigenous people were ruined, and the educated young people despaired [8].

The "sustainable" practices of indigenous forest dwellers are usually transmitted orally, but in India and China there is also a considerable amount of literary evidence for the awareness of the need for human self-restraint in the use of natural resources. Often such references have to be "decoded" in order to be understood. Thus, for instance, ancient Indian prescriptions of the type of wood to be used for the great sacrificial fires appear to be unnecessarily detailed, but in practice such prescriptions prevented indiscriminate logging near the fireplace and induced a careful harvesting of branches and trees. Similarly many ancient Chinese texts, particularly those in the Taoist tradition, reflect a respect for the balance of nature. Steeped in magic thought, they also require "decoding".

Ancient wisdom and the experiences of forest dwellers were displaced by the advance of settled agriculture. The old Indian saying "the land belongs to him who first cleared it" reflects the progress of the settled peasant whose activities certainly amount to a massive interference with nature. The saying also points to the institution of permanent ownership which was unknown to swidden cultivators. Ownership implies an instrumental relation to the land, it also encourages the peasant to think of improving it so as not only to sustain but to enhance its productive capacity. Due to the vagaries of rainfall, the settled peasant had to look for artificial irrigation and/or the adjustment of his crops to the prevailing climatic conditions. In many regions which permitted only the cultivation of crops requiring little water, peasants mixed crops so as to spread their risks. But rice, the grain which guarantees the highest yield but also requires much water, is conducive to the spread of a monoculture, often producing more than one harvest per year. Rice responds very favorably to higher labour inputs. This produces a dynamic relationship: where rice can grow, more people will also grow and they in turn can grow more rice with added labour. Thus rice induces a massive interference of man with his environment and encourages the complete clearing of all land suitable for its cultivation. This development is "sustainable", but it implies that rice smothers all other types of cultivation like a wet blanket wherever climatic conditions permit its growth.

„Sustainability" would thus have a very different meaning for a swidden cultivator and a peasant engaged in advanced wet rice cultivation. While the swidden cultivator would gear his activities to a minimal interference with nature, the rice grower would aim at improving his plot of land in order to get the most out of it. The mindset would change with the environment. The swidden cultivator would live in a landscape where he would hardly see any traces of human interference, while the rice growing peasant would be used to seeing wide expanses of green fields maintained by intensive human labour.

Rapid population growth and the rise of the modern state in India and China then led to a further change of the mindset. The governments of such states felt that they could sustain their large population only by dramatic increases in productivity. One way of doing this was by relieving the pressure on the land through industrialisation, but a more immediate way was that of improving the productivity of agriculture. Irrigation by means of large dams which would also yield a substantial supply of energy was a measure which was highly valued by these technocratic governments. Unfortunately such governments, though claiming to adhere to scientific standards, were often carried away by singleminded devotion to grand schemes. Mao's support for the construction of a huge dam across the Yellow River is as case in point. A Chinese engineer who accurately predicted that this dam would be useless as it would silt up very quickly was not listened to but punished. The same engineer also warned against the construction of the Three Gorges Dam on the Yangtse. But whereas his predictions concerning the Yellow River dam came true within a short time, it still has to be seen what happens to the Three Gorges Dam [9].

The Indian government has also shown great faith in dam building. The silting up of dams has been witnessed in India, too. Nevertheless the government has adopted ever more ambitious schemes, the most recent one being the controversial Narmada Dam. Public attention was mostly directed at the fate of the people whose land would be submerged. There was hardly any debate about the threat of earthquakes. The tectonic history of India has been determined by the huge shelf which threw up the Himalaya by moving to the North at a rate which is quite rapid as compared to other movements of this kind. Usually earthquakes have occurred at the Northern rim of this shelf. But the earthquakes at Latur (Maharashtra) in 1993 and at Bhuj (Gujarat) in 2001 have shown that by now the center as well as the Western rim of this shelf are prone to tectonic upheavals. Bhuj is ca. 400 km to the Northwest of the dam, and Latur ca. 450 km to the Southeast of it. The "sustainability" of the dam could thus be a major problem.

Applying the term "sustainability" to the ambitious projects of modern technocratic governments may sound farfetched, if we remember what German foresters had in mind when they coined this term. But the potential of human interference with the environment has progressed very dramatically, indeed, from the minimal operations of the swidden cultivator to the projects of dambuilders. Actually the building of huge dams is only the most visible form of this interference. There are many other forms of such modern interference whose "sustainability" is rather doubtful. Moreover, there is the general problem of unintended consequences. A striking example of this kind has been desribed in this volume (see Chapter 7). The planners of water management of California acted on the principle that water flowing into the sea is lost to agriculture. They stopped this flow and agriculture expanded. At the time they made their plans they could not forsee that modern agriculture works with massive doses of pesticides which contaminate the water which now no longer runs off to the sea as it used to do in earlier times. The human mind is wedded to the idea of linear progress and this applies with a vengeance to modern technocratic governments whose planners get credit for the visible success of their projects. "Sustainability", however, has to be assessed in terms of multiple causation and a variety of future scenarios. Mankind can no longer return to the world of the swidden cultivator who lives in harmony with nature and protects biodiversity in sacred groves.

Thus the modern governments of India and China cannot be expected to turn to traditional ideas of "sustainability", but when they claim that their ambitious projects are based on "scientific" planning, they should make use of all sources of knowledge and not limit their vision by means of technocratic blinkers.

Mahatma Gandhi compared the single minded pursuit of technological progress to the moth which circles around the flame until it is consumed by it [10]. Unfortunately his praise for the India of the villages and his condemnation of large-scale industry have often been interpreted as a kind of retrograde romanticism totally irrelevant for the modern world. Actually he was concerned with "sustainability" in the broadest sense of the term. The words he used were sometimes inadequate for conveying the message that he wanted to get across. His "experiments with truth" as he called his actions often reflected his ideas more accurately. The quest for sustainability should be regarded as such an "experiment with truth". Just like Gandhi's truth, sustainability cannot be taken for granted, but must be tested in action.

The Gandhian design of nonviolent action can also be applied to the movements of resistance to the measures of technocratic regimes. This was demonstrated by the Indian Chipko-movement, whose members prevented the felling of trees by embracing them [11]. Such actions are, of course, possible only if the regime concerned is not prepared to repress them brutally. It is doubtful whether a similar movement could have operated in Mao's China in the case mentioned above. Contests concerning sustainability thus depend on the respect for civic rights. If a technocratic government is also autocratic it can define sustainability so as to suit its own plans. The respect for civic rights may thus be more important for the maintenance of sustainable development than the recourse to ancient wisdom.

12.4   Sustainable Societies in Asia

The relation between civic rights and sustainable development may explain why in recent times the concept of sustainability has been extended so as to encompass human societies. Actually the term "sustainable society" has been included in the mission statement of the European Academy of Sciences and Arts, but no definition of the term has been provided in that statement. We may assume that this concept would not just refer to specific measures aimed at guaranteeing sustainability but to the entire social setting required for sustainable development. A society which recklessly devastates its habitat and depletes its resources would be an "unsustainable" one. The society of swidden cultivators mentioned above would be a sustainable one as long as it follows the practices which have been described. But how to evaluate larger societies such as the Indian or Chinese ones in these terms? The processes of environmental degradation which concern these societies are so vast that individuals cannot grasp them at all. They have to trust in the rule-making capacity of the state which imposes restrictions on "unsustainable" practices.

The Republic of India and the People's Republic of China have produced a large amount of legislation in this field, but the implementation of the respective rules remains a problem. Very often compliance with such rules means additional costs which cannot be passed on to the consumer by means of an increase in prices unless all producers follow them without any exceptions. Inspection is often inadequate and inspectors can be bribed. A recent case study of industries in Western India has shown that the treatment of effluents prescribed by law was neglected for these reasons. Either the required equipment had not been installed at all or it was not properly utilized [12]. This study referred to private sector enterprises. Unfortunately the record of Indian public sector enterprises is also disappointing in this respect. This means that the state does not take care to implement its rules in enterprises which it owns [13]. If this kind of negligence continues, the consequences will be very severe. Fortunately the Indian courts have been active in enforcing environmental regulations. In one instance they ordered the closure of several hundred tanneries in the state of Tamil Nadu, because the effluents of these tanneries polluted the rivers. Joint efforts of two major national laboratories then helped to solve the problem of treating effluents and the tanneries could resume their production. This is an encouraging success story. But it normally takes some time for the courts to swing into action. Moreover, the courts have to be moved by citizens who take an interest in such matters. Non-governmental organisations (NGOs) often play a decisive role in this context. In fact, if one would wish to define indicators of a sustainable society one could list the actions taken by NGOs in defending sustainable development.

The organs of the state often rest content with legislation when has been put on the statute books, but it is only when such legislation is enforced and tested in a court of law that it becomes of real value. Whereas legal action of various kinds may be a good indicator for social concern with sustainability, this may not be enough to define the quality of a sustainable society. There are many issues which are of importance but which may not be amenable to legislation and judicial decisions. The discussion of matters relating to sustainable development in a free press and other media may be a suitable indicator for a wider social concern with such issues. Similarly the attention paid to sustainable development in school and university curricula would indicate "institutional" social awareness in this field.

Empirical studies along these lines would probably show that Asian societies do not yet qualify for being called "sustainable societies". They are also subject to Western influences with regard to the standards of consumption which are clearly "unsustainable" as far as they are concerned. A simple example may illustrate this: Germany has at present about one car per 1.5 citizens; China and India have together a population of about 2.4 billion people and should therefore be entitled to altogether 1.6 billion cars. If they would actually claim this entitlement, it would be a nightmare. But no German could blame them for wishing to attain his standard. This example would lead us to the conclusion that the sustainability of societies cannot be judged in isolation. If we adopt uniform standards of consumption for all societies we would arrive at an unsustainable world. On the other hand, it would be impossible to reduce the standards of Western societies to those of the Asian ones. In India about 4.2 million cars (excluding trucks and busses) were registered in 1995, i.e there were 238 Indians per car. If we would apply this standard to any Western country, there would be very few cars around. In 1980 there were only about 1 million cars in India, numbers had thus increased by four times in 15 years [14]. The scope for a further increase is considerable. But how to limit the dimensions of consumption? We do have a World Trade Organisation (WTO) which tries to widen the scope of free trade, but we do not have a World Sustainability Organisation (WSO). The fate of the Kyoto Protocol shows that even limited and very specific initiatives of this kind are not acceptable. A truly effective WSO would probably come into conflict with the WTO which promotes increasing economic growth based on free trade. The WTO would even endorse the trade in rights to emissions etc. Rich nations already buy such entitlements from poor countries. This would enhance the differences between rich and poor nations rather than help to reduce them. A WSO should certainly not encourage this, but for that very reason it will not be established.

The conflict between the rich and the poor, which is an international one in the context just mentioned, is a domestic problem for China and India. The "poverty line" is difficult to define, some statistics are based on nutrition, others include a variety of basic amenities such as shelter, access to drinking water etc. If such additional factors are taken into consideration, at least one third of India's population lives below the poverty line. Chinese statistics are not very transparent in this respect, but poverty is widespread there, too, particularly in the Western provinces. To the extent that the poor millions of these two giant countries attain a better standard of living the problem of sustainable development becomes more obvious. It sounds cynical if one maintains that Asian societies are sustainable as long as large parts of their populations rest content with low living standards, but unfortunately this is true. The political systems of these countries face a serious dilemma. They are judged in terms of their ability to reduce poverty and if they are successful in this endeavour, they create problems which will put them to severe tests. The "revolution of rising expectations" which has been discussed a long time ago, has been delayed, but now it seems to be in full swing. These expectations encompass higher living standards and increasing consumption, they are not aimed at sustainable development. It will take a great deal of educational effort to convince people of the fact that the fulfilment of their expectations depends on sustainable development. But this cannot be done by telling the poor that they must remain poor so that the rich can sustain their standard of living. China and India will have to face this problem within their own national boundaries and it remains to be seen how they cope with it.

12.5   Conclusions

In another contribution to this volume (Chapter 13) "assertive citizenship" has been highlighted as a main support of sustainability. This is certainly true, but this kind of assertiveness depends on motivation and information. If threats to sustainability are not recognised nobody will take action against them. Most of these threats are creeping ones, they do not claim everyone's attention by an awful visibility like the destruction of the World Trade Center. Moreover, they are very complex (see Chapter 6) and require careful analysis. The results of such an analysis are difficult to communicate, they must be "translated" into messages which can be understood by ordinary citizens – including the people of Asia. This will be a major task, it should be taken up by the World Sustainability Organisation (WSO) which has been mentioned above. Under present political conditions such a WSO would have no chance to get established as a regulating agency, but it would perhaps be supported if it were only a monitoring agency disseminating information. Citizens could then base their actions on this kind of information. It would also become an essential part of education and influence the minds of the younger generations. They would certainly want to know what kind of world they are going to inherit.

References

[1]   S. K. Parikh in India Development Report (Ed. S. K. Parikh) 1999–2000, New Delhi, pp. 65.

[2]   D. Rothermund in Das Bevölkerungswachstum", Indien, Ein Handbuch (Ed. D. Rothermund) München, 1995, 59 f.

[3]   J. Shapiro, Mao's War Against Nature, Politics and the Environment in Revolutionary China, Cambridge, 2001, p. 29.

[4]   Centre for Science and Environment, The State of India's Environment 1982, A Citizens Report, New Delhi, 1982.

[5]   G. Spreitzhofer in "Gunst und Ungunsträume in Südostasien" (Eds. P. Feldbauer/ K. Husa/R. Korff), Südostasien, Wien, 2003, p. 103.

[6]   K. Husa, /H. Wohlschlägl in "Südostasiens 'demographischer Übergang'. Bevölkerungsdynamik, Bevölkerungsverteilung und demographische Prozesse im 20. Jahrhundert" (Eds. P. Feldbauer/K. Husa/R. Korff), Südostasien, Wien, 2003, p. 133 f.

[7]   G. Spreitzhofer in "Gunst und Ungunsträume in Südostasien" (Eds. P. Feldbauer/ K. Husa/R. Korff), Südostasien, Wien, 2003, p. 89.

[8]   J. Shapiro, Mao's War Against Nature, Politics and the Environment in Revolutionary China, Cambridge, 2001, p. 173 f.

[9]   J. Shapiro, Mao's War Against Nature, Politics and the Environment in Revolutionary China, Cambridge, 2001, p. 52–58.

[10]   D. Rothermund in Mahatma Gandhi, Der Revolutionär der Gewaltlosigkeit, Eine politische Biographie, München, 1989, p. 400.

[11]   Centre for Science and Environment, 1982, p. 42 f.

[12]   Y. Joshi in "Policies for Sustainable Industrial Development and Their Implementation in Western India" (Ed. D. P. Sinha), South Asian Management: Challenges in the New Millennium, Hyderabad, 2002.

[13]   S. K. Parikh in India Development Report (Ed. S. K. Parikh) 1999–2000, New Delhi, p. 90.

[14]   S. K. Parikh in India Development Report (Ed. S. K. Parikh) 1999–2000, New Delhi, p. 300.

13 Sustainability of Development and Valuation of Non-renewable Resources: An Analysis in the Context of Local Cultures

Michael von Hauff*, Amitabh Kundu**

* University of Kaiserslautern, Gottlieb Daimler Str. 42/404, 67663 Kaiserslautern, Germany, hauff@wiwi.uni-kl.de
** Jawaharlal Nehru University, Center of Studies of Regional Development, New Delhi 67, India, amit0304@mail.jnv.ac.in

13.1 Introduction

Increasing unease with the modern era of development in large parts of the world, characterised by undifferentiated obsession with technology, consumerism and modernity, is expected to open up new paths more sensitive to values and local cultures. It is now widely recognised that strides in science and technology, emphasis on economic efficiency, reliance on management skills etc. by themselves are incapable of solving the long-term problems facing humanity. There seems to be a growing conviction that an appropriate long-term development strategy can be chalked out only by incorporating socio-cultural parameters as an integral part of the framework.

Researchers, planners and administrators in developing countries have often pleaded for bringing cultural norms into the heart of the current policy debates, particularly those pertaining to environmental degradation. A number of policy documents issued by national governments underline the importance of local culture in preservation of environment. Experiments conducted by several NGOs reveal that local culture and the process of economic development can enrich and enhance each other. Despite all this, development strategies in few Third World countries reflect a desired level of sensitivity to cultural norms and indigenous values or incorporate these in operationalising policies and programs, to achieve sustainable development.

A section of world political leaders and development analysts, working on socio-cultural aspects, have voiced their concern and apprehension that in several countries of Asia, Africa and Latin America, there exist "toxic cultures" that handicap the process of sustainable development. A study overviewing the development experience of Latin American countries during seventies and early eighties [1] concludes that their different

Global Sustainability. Edited by P. A. Wilderer, E. D. Schroeder, H. Kopp
Copyright © 2005 WILEY-VCH Verlag GmbH & Co. KGaA, Weinheim
ISBN: 3-527-31236-6

levels of performance have been determined by the proximity of their culture and value system to those in North America. Similarly, Landes [2] argues that cultural characteristics must be judged by the extent to which these fulfil the "duty" of "keeping up" with their northern neighbours. Casual empiricism and quick generalisations of this type would be more a hindrance than help in understanding the relationship between socio-cultural parameters and economic development on the ground. It is extremely important that this issue is examined with empirical rigour, focussing on a few critical sectors and analysing the relevant programmes, policies and projects in the context of local culture and practices.

We have to ask to what extent the cultural characteristics of the countries in the Third World can be defended and what are the norms and practices that ought to be discarded. More importantly, we must ascertain the standards or criteria through which these characteristics are to be evaluated. It would also be worthwhile to investigate how the warning given by the World Bank against not incorporating "the social knowledge into financially induced growth programmes" has impacted on the national and international organizations, including the Bank itself, in their policy making. Indeed, this needs to be done through rigorous empirical investigation into the changing nature of the projects and the criteria adopted in their evaluation. Further, it would be important to explore the possibility of using people's culture for their own empowerment, leading to assertive citizenship, so that a socio political environment is created that can promote policies and actions for sustainable development.

Keeping in view the above perspective, the present paper begins by analysing the concept of sustainability in economic theory, focussing on certain methodological issues relating to valuation of non-renewable resources. This has been done in the second section, which follows the present introductory section. The third section narrows down to focus on technological choices in energy sector and discusses how incorporation of cultural values in the development process and public decision making can help in evolving a development strategy ensuring sustainable resource use. The fourth section overviews the recent efforts to incorporate cultural dimension in the development projects at national and international levels and discusses their implications for sustainability. Additionally, it underlines the need to multiply such initiatives and examines how the norms and values embedded in local culture can be harnessed in creating the right kind of pressure on the process of socio political decision making.

13.2   Sustainability in the Context of Economics of Ecology

It is difficult to envisage that governments in less developed, as well as developed, countries relentlessly pursuing the dictates of globalization and economic liberalization would be able to derive policies and programs that impose a discipline on market forces so that the interests of future generations are not compromised. In this context, it would be interesting to begin by reviewing the positions taken by various theories of economic development with regard to sustainability. More specifically, it is of importance to know how non-renewable resources have been valued so as to ascertain what kind of financial discipline needs to be imposed on the rates of their extraction and nature of application.

The principle of sustainable development has been interpreted in many different ways, each interpretation giving rise to different concepts and strategies. Frequently, however, these start from the same point, namely the definition given in the Brundtland Report [3]. The report describes the path towards sustainable development as that which "meets the needs of the present without compromising the ability of future generations to meet their own needs". The definition clearly shows that the principle of sustainable development is firmly rooted in value judgements requiring assessment of needs and abilities of the people and their intergenerational comparability.

The definition is thus normative in character. It makes a normative demand or invokes decisions that call for equal opportunity of development for all people living at present times and those coming in future years. The danger in adopting this kind of general definition of the term is that it opens up the possibility of alternate interpretations, which are often imprecise. It is, therefore, unsuitable both for theoretical analysis and for policy linked research. This problem has been discussed in great detail by Blättel-Mink [4]. Principally, the Brundtland Report puts forward demands in two directions.

- First, it requires the needs of the present generation to be met in an equitable manner. More specifically, this involves equitable settlement of interests of people in industrialized countries and those in the developing world. It is a question of intragenerational equity.
- The second demand is aimed at preventing the lifestyle of present generations from impairing the chances of future generations to meet their needs. The aim here is intergenerational equity, requiring the process of economic development to allow the present and future generations to utilize the world's resources in an equitable manner.

The above discussion underlines the need to impose stringent constraints on the development strategies of individual countries, following the principles of sustainable development. The constraints or conditionalities have become binding over the years because of the limited assimilative capacity of the environment and exhaustible nature of nonrenewable resources, on which pressures are increasing enormously. It is, for example, evident that the development path and associated consumption pattern adopted by Europe and North America are unsustainable. In no case, this pattern can be applied on a global scale because of its high demand on non-renewable resources and the problems of environmental pollution implicit in this model of growth.

It is here that the dilemma confronting the present generation becomes manifest. Increased and improved provision of the means of satisfying basic needs of the large parts of the population (particularly those living in poverty) in developing countries presupposes an increased access to productive assets, which can only be achieved by speeding up economic growth. This, given the technological options available to these countries, is bound to put strains on environment. The only way of improving economic growth in an ecologically compatible manner would, then be by, one, removing constraints of capital and technology faced by less developed countries and, two, by providing them access to world market so that they do not have to over exploit their nonrenewable resources. And, for operationalizing both these components of the strategy, the industrialized countries would have to make sacrifices in terms of level of income growth or energy intake [5].

Are the industrialized countries willing to accept a deceleration in economic growth as a result of self imposed restrictions on the demand side in addition to bringing about improvements in the efficiency of production by employing more energy efficient technology? The challenge involved reflects how difficult it is to move away from the extremely unsustainable baseline situation we presently have in industrialized countries in particular (but also increasingly in the developing countries) to a course determined by sustainability. It has to be assumed that this transformation will not be easy and will require serious socio-political realignment.

For a transition to a sustainable growth path, the issue is not merely of whether non-renewable resources can be replaced by renewable ones. In the context of the satisfaction of future generations' needs, one frequent topic of debate is whether human-made capital (i.e. real capital) can substitute for natural capital. If we assume that the two are by and large substitutable, and that future generations will accept this substitution, we arrive at the concept of weak sustainability. The assumption is that the future generations do not care whether capital is real or natural [6]. Following this approach, a reduction in reserve of resources, say minerals, can be compensated through an increase in the stock of infrastructural facilities like roads and rails.

Scholars such as Daly [7] have critically examined the way neo-classical environmental economists view sustainable development. He observes that natural capital is one-way flow of inputs into economic process. Unless a way is found of replenishing this source, the inflow would have a limited time horizon. This implies that real capital can never replace the inflow or its source. The inescapable conclusion which emerges from this is that real capital cannot be a substitute for natural capital, despite there being several complementarities between them [7]. Further, it also has to be kept in mind that the environment is not solely a supplier of inputs for the economic process. Besides providing a life-support system to population and all their pursuits, it is also the *sine qua non* of economic and social activities [8].

Importantly, the possibility of technical substitution in the real world is frequently limited, and can by no means be extended indefinitely. This could result in irreversible destruction of natural capital, such as reduction of natural diversity, destruction of ecological systems etc. that can not be replenished through production of man made capital. In other words, weak sustainability position offers no basis for a clean-cut turnaround to sustainability.

Understandably, ecologists of a pristine pure variety have often opposed the neoclassical position regarding substitutability, noted above. They fundamentally question the possibility of replacement of natural capital by real capital and call for the strict preservation of the former for the sake of environment. This can be justified if, using the argument derived from the "cake-eating problem", ecological sustainability is given the utmost priority. The derived argument would then be that no generation could touch or modify natural resources. The result would be that all generations would have to forego utilization, but this in its turn would lead to a kind of prisoners' dilemma and result in totally inefficient and unacceptable solutions for mankind.

As a rule, therefore, the starting point must be a willingness to regard utilisation of resources and sustainability not as contrary aims, but as two components of a comprehensive development path. For example, the proposition that the manufacturing industry in both advanced and less developed economies is economically indispensable if the needs of the population are to be met, despite their certain negative consequences in terms of ecology, would be acceptable. Seen from this perspective, the "pure" ecologists' position is untenable. This is because it is not possible to envisage, given the current level of technology, an industrial production process that does not need natural resources as inputs and does not have to deal with the problems of complementary products, emanating in the form of pollutants [9].

Happily, sustainability has generally been interpreted to imply not compromising the ability of future generations' to meet their own needs rather than maintaining the physical resources in their naturally pure state. One can argue that future generations would possibly have lesser volumes of physical resources due to exploitation natural goods etc. but that can be adequately compensated through increased availability of real capital and greater technical expertise in dealing with natural resources. This would allow natural goods like mineral resources, water, air and soil to be used in the production process but within a framework of increased ecological efficiency.

It is expected that progress in environmental technology will be able to reduce the burden on environment to only a limited degree. The crucial question would therefore be whether this technology can develop or be encouraged to advance quickly enough to prevent the occurrence of irreversible environmental damage with its obvious consequences such as global warming, etc. Although progress in environmental technology has been very rapid in recent years, especially in terms of energy-saving systems, the spread and use of these advances has been relatively low.

There is of course the danger that the present generation will use its own calculation of welfare to interpret and determine the basic needs of future generations. As demanded by the Brundtland Report, this should be avoided. The objective should be bequeathing to future generations a stock of natural resources that is as undepleted as possible, together with an environment that has a commensurate assimilative capacity. It is in the context of operationalizing such a strategy that the major shortcomings of the theories of sustainable development become clear. This is because, little attention has been paid to the question of whether sustainability is a condition imposed by technology, by biologically determined minimum standards, or as a set of preferences for sustainability. Unfortunately, most studies have concentrated on minimum standards and ignored the socioeconomic processes that determine the actual choices in the field.

The analysis of sustainability as being linked to societal preferences (welfare function) assumes that every society makes a choice between alternative paths of welfare maximisation [10]. In this context, special attention deserves to be given to Rawls's theory of justice [11]. If his theory is transferred to inter-generational equity, it can be argued that the "veil of ignorance" will force all individuals or generations to strive to safeguard the existence of future generations. The major criticism of the approach is that the preferences of the present generation would be neglected for the sake of future generations (tyranny of the future). The utilitarian rule comes to the diametrically opposite conclusion as it can argue that with increasing distance, the present generation will give lesser importance to the utility of future generations, and will finally give it a value of zero (tyranny of the present). In other words, the utilitarian rule does not fulfil the criterion of sustainability.

It should be possible to overcome the tyranny of the present and the future by combining the two criteria noted above [10]. According to "representation theory", a welfare function can be derived that represents sustainable preferences [12]. The welfare function for sustainability would, thus, comprise two part functions, such that both present and future are accorded significance, with a built in trade off between the two, reflecting societal preferences for sustainability.

Apart from providing a sounder theoretical base, an analytical distinction between social preferences and expectations about technological and biological standards has a distinct advantage in the context of designing environment policy. Social consensus with respect to sustainable preferences is an important condition for an appropriate assessment of the role of technological and biological developments. While different groups within a society may value the latter differently, emergence of common sustainability preferences will facilitate formulation of a consistent environment policy.

In principle, this will render irrelevant the controversy on the issue whether real capital can be substituted for natural capital. The main difference between ecological economics and the neo-classical strain of environmental economics is that the former makes the economic system subordinate to the ecological system [13, 14], which may be objected to by the latter. The core concern of this relatively new discipline, however, should be to provide justification for and facilitate implementation of the strategy of sustainable development. This can be done within the framework proposed above, which happily is being increasingly accepted and acknowledged since the end of the 1980s (for more details see Rogall [15]).

Proponents of ecological economics have thus come to accept a limited interchangeability of natural, real and human capital. However, this is based on the maxim of the fundamental preservation of the vitality of the natural environment. As seen by ecological economics, the principle of sustainable development requires rules for decision making and action must be made more precise and put into practice. The three rules depicted in Figure 13.1 should be highlighted (see Brundtland Report [11], Pearce/Turner [16], Cansier [17]):

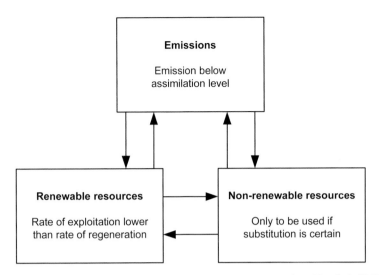

Figure 13.1 Rules of action for sustainable development Source: based on Vornholz [26]

The assimilation rule and the use of renewable resources relate to the ecosystem's biological, chemical and physical powers of regeneration. These stipulate that environmental media viz. water, air and soil, must only be exploited to the extent that they can regenerate naturally. In this way, pollution beyond certain limits and non-repairable damage to environment can be prevented. The same applies to renewable resources as well. As long as they are not used up faster than their natural rate of growth, depletion of stocks will not be a problem. For non-renewable resources, it is unfortunately not possible to formulate a scientifically sound rule to put a ceiling on use. If oil, natural gas and other non-renewable resources are consumed, stocks cannot be kept constant. For the reasons mentioned, however, it does not appear to make sense to forego them permanently.

When looking more closely, it must said that even the first two above-mentioned rules of action are relatively vague [17]. Let us look first at the rule for renewable resources. When exploiting forests, a balance between the wood taken out and the right amount of replanting is not the only issue. As an ecosystem, the forest can fulfil its many functions – production, protection and especially recreation – in the best possible way only when ecological forest management systems are introduced.

Assimilation capacity is determined by the natural ability of self-cleansing. This is a matter of the decomposition of pollutants in water by microorganisms and self-cleansing of air by chemical and physical processes. Understandably, short-term decomposition of pollutants is not a problem. The situation, however, would be difficult if pollutants accumulate in environmental media over a long period. The greenhouse gas carbon dioxide, for example, remains active in the atmosphere for more than 100 years [17]. The decisive factor, therefore, is that the tolerance of nature should not be taxed over a period lasting several generations. The danger here is that the absence of definite figures will lead to wrong decisions, giving rise to situations that are ecologically irreversible.

The three classical rules of action can be extended to include a few more economic principles [18]. For example, a case can be made to impose scarcity rents for exploitation of non-renewable resources on the users. In addition, institutional arrangements should be made to ensure that resources are not over used and used more efficiently by employing technologies that are kinder to environment. The preservation of biodiversity can also be incorporated in the arrangements, as a rule of action. These rules will not be discussed in more detail here, as these have been derived from the classical rules of action, noted above.

The above rules along with the derived stipulations help in formalising the framework for working out the societal preferences for environmental policies. The rules are extremely demanding and require structural changes in the system of formulation and implementation of development strategy. These would also necessitate redefining the needs of the present generation that must be satisfied as also rethinking on the methodology for measuring welfare for inter temporal comparison (on this topic, see Martinez-Alier, Munda, O'Neill [19]).

13.3 Energy Options and Sustainable Development in the Developing World

Despite recognition of the issue of sustainable development in international documents, declarations etc. and the urgency of launching corrective actions and programs, there is no indication that the countries in general have chosen this option or made significant advances in moving towards that direction. One of the reasons cited in this connection is the contradiction between the concept of environment evolving through culture, scripture and religious thinking, particularly in the developing countries, versus being driven by the market or present policies of the government. It would be useful to analyse and understand the reasons for this contradiction, taking the energy sector as an example and assess its implications in terms of the environmental crisis facing the World.

The growth of fossil fuel based energy system over the last century has led to increasing concentration of oil in the hands of a few multinationals. It has also led to sharp disparity in energy use, and ultimately social well being, among the people in different countries of the World. This can be illustrated using the figures of the Worldwatch Institute. The data reveal that the richest 20 percent of the population consumes over 58% of the world energy while the bottom 20% consumes less than 5%. Besides such glaring disparities, the enormous cost of the fossil fuel in terms of concentration of the green house gases pose a serious threat to human race.

A recent report from the UN Security General's Office [20] on the causes of conflict and promotion of durable peace and sustainable development in Africa proposes to hold a conference focussing on energy security. The underlying premise of the Report is that the next century must ensure that the current reliance of the world population on fossil fuels shifts toward systems that employ more sustainable and renewable forms of energy. The conditions are ripe for this shift in developing countries given their generous endowment of renewable energy resources and the fact that they are can leapfrog out-dated 20th century technologies and acquire energy systems that are more environmentally sustainable at a low cost.

It is important to observe that two of the last-mentioned rules of ecological sustainability, discussed in the preceeding section, are extremely relevant in the context of the energy sector. This can be demonstrated clearly for developing countries like India. Here, pollution caused by the energy sector is very high due to the nature and quality of the fuel used. Attention has been drawn to the coal-fired power stations that provide 60% of the total generation capacity. In addition, the coal mines in India produce a large amount of soot giving rise to high levels of pollution. A review of the macro policies suggests that this is unlikely to change appreciably in near future.

Going by the effective demand, backed up by affordability (as opposed to normative requirements), there is an energy shortage of approximately 11.5% in India, which in peak periods becomes as high as 18%. Contrastingly, the country has a huge renewable-energy potential (especially solar and wind power), of which only a very small proportion has been exploited. As a consequence of these inoptimal policy choices with regard to source of energy, technology of production, system of distribution, and pattern of demand, present-day India is facing serious environmental crisis.

13.4 Cultural Norms and Concerns for Sustainability in Formulating Development Strategy by National and International Agencies

A genuine concern for incorporating social and cultural values as important elements in the development strategy can be noted in policy documents released by several international organisations and and many have made concerted efforts in recent years to move towards that objective. In 1991, the World Bank brought out a paper [21] demonstrating that, of the 57 projects financed by the Bank, those having socio cultural compatibility with the local situations registered rates of return twice as high as that of non compatible projects. Its implicit suggestion is that the national governments and international agencies should formulate development projects based on in-depth analysis of local socio cultural situation.

In 1995, UNESCO published a Conference report jointly with the World Commission on Culture and Development entitled "Our Creative Diversity", highlighting the importance of different cultures and their distinct place in the process of economic development. Inaugurating the Conference, Wolfenson, the President of the World Bank had observed that irrespective of how "you define culture, it is increasingly clear that those of us in the field of sustainable development ignore it at our peril" [22].

The significant development in this context is the agreement between World Bank and leading religious leaders of the world in 1998 for setting up a Working Group to launch projects to fight hunger and preserve cultural heritage. Subsequently, the Millennium Summit was held in 2000, which had sustainability as one of the thrust areas. It underlined the significance of the commitment to respect local cultures and values by different governments and development agencies as precondition for success of their development strategy. UNDP [23], too, drew attention to a "A new Generation of Poverty Programs" that seek to involve community organisations for directly articulating needs and priorities of the people.

More recent documents from the World Bank like the World Development Report 2000-01[24] and Voices of the Poor [25] underline similar concerns and plead for giving greater attention to culture related issues. The World Development Report concludes that "history shows that uniformity is undesirable and that development is determined to a great extent by local conditions, including social institutions, social capability ..."

Importantly, World Faith Development Dialogue in 2001 tried to identify key issues in working out a development strategy, keeping in view the contemporary socio-political situation in the world. It tried to derive its conclusions using nine sets of papers written by scholars of nine different Faiths. Based on these, it advanced a critique of the contemporary development strategy advocated by World Bank. It argued that the perspective put forward through World Development Report is giving wrong signals to the people and the governments, across the countries of the World. The Dialogue urged international development organisations to incorporate spiritual dimensions within their development programmes and sought to provide assistance towards this goal.

The relationship between the Baha'i International Community and the UN organisations needs a special mention here. The Community became affiliated to UN Enviromental Programme in 1976 and obtained consultative status with UNICEF in 1980. Baha'i have been taking up UN sponsored projects in different countries of the World, their number going up to about 3000. The Community has been called upon to play important role in Pre Coms for various International Summits and has played a role in drafting the Universal Code of Environmental Conduct (1990), International Legisltion for Environment and Development (1991), The Earth Charter (1991) that have a direct bearing on sustainable development.

It is, however, unfortunate that contemporary studies on culture of different communities are not being conducted with the objective of achieving sustainable development but have been motivated by other goals. It is rare that one finds planners and administrators in international organisations showing the sensitivity and keenness to understand cultural aspects in developing countries in an attempt to integrate the local "communities into programmes designed in another context by people of another culture". Even in cases when this is done, an attempt is made to make an instrumental use of culture which excludes the possibility of any genuine empathy and relationship of mutual learning between the development worker and the people in the community.

Formulation of programs based on local culture would require openness to spiritual and religious concerns and an understanding that development can not be promoted through technical skills alone. The criteria of success of the programs, therefore, should not be materialistic but people based. These would envisage learning from the experience, knowledge and outlook of traditional and religious leaders in the community.

It is only then that it would be possible to have them as effective multipliers, taking advantage of the authority they enjoy within the community. Certain training programs can be organised for them and other community leaders so that they can work with the people to strengthen their capacity to manage their economic lives in consonance with cultural values. This would force development agencies to look beyond the dualistic approach which separates spirit from matter, culture from economics and ethics from growth.

13.5 Conclusions

The issue of why the capitalist path of development seems to succeed in the Western world but fails in most other countries has led researchers to examine the importance of local cultures. A view has emerged that the cultural characteristics in the less developed world impose constraints and can slow down or even reverse the growth process. A section among researchers and policy makers have, thus, hastened to judge the efficiency or acceptability of cultures in terms of their compatibility with those of western countries. The paper argues that casual empiricism underlying such generalizations come in the way of a proper understanding of the relationship between socio-cultural parameters and economic development.

It demonstrates that there is an urgent need for empirical identification of the cultural characteristics in the countries of the Third World that need to be defended or even be strengthened for promoting sustainable development. More importantly, it would be important to come to certain consensus on the standards or criteria for evaluation of these characteristics and ensure that social knowledge, which is more often than not, in conformity with principles of sustainability, is incorporated in the financially induced growth programs and other spheres of policy making.

Many of the less developed, as well as developed countries, pursuing the dictates of globalization would have serious difficulties in agreeing on policies based on observance of these principles. Indeed that would necessitate imposition of certain kind of discipline on market forces and growth programs. A analysis of the theories or models of economic development in the context of the principles of sustainability suggests that countries must adhere to strong normative principles or classical rules of action. More specifically, the methodology for valuation of non-renewable resources, implicit in these models and their derivatives, requires that their extraction must remain within well defined limits so that no irreversible damage is done to the environment.

It is important that these rules are used in formalizing the societal preferences for environmental policies. These would necessitate redefining the needs of the present generation or placing ceilings on these, as also rethinking on the methodology for measuring welfare for inter temporal comparison. Imposition of scarcity rents for exploitation of non-renewable resources, ushering in administrative and legal reforms to ensure that resources are not over used and that eco friendly technologies are employed, would be part of a new institutional arrangement. The preservation of biodiversity can explicitly be incorporated in the arrangements, as an important rule of action.

Unfortunately, despite the issue of sustainable development figuring in various national and international documents and declarations, the countries have rarely chosen this option or made significant advances towards that objective. One of the reasons cited in this connection is the contradiction between the concept of environment evolving through culture, scripture and religious thinking, particularly in the developing countries, and that emanating from the theories of sustainable development.

An analysis of existing literature suggests local culture could have a greater impact in promoting sustainable development than the current programs being launched by the governments or the prcess driven by the policies of globalization. Reviews of development projects reveals that those having socio cultural compatibility with the local situation have been far more successful than the non compatible projects. It would, therefore, be desirable for national governments and international agencies to formulate development projects, taking into consideration the socio-cultural specificity of the local situation, purely from the viewpoint of long-term economic efficiency. There is, thus, an urgent need to have openness to spiritual and religious concerns and seriously understand and incorporate, to the extent possible, the outlook of traditional leaders in the community, through systematic research and negotiation. It is only then that it would be possible to get away from the dualism between spirit and matter, between pursuit of cultural values and materialistic goals and be on the path of sustainable development, ensuring benefits to all sections of society.

References

[1]   van den Bergh J. J. C. M., Ecological Economics: Themes, Approaches, and Differences with Environmental Economics, Discussion Papers Tinbergen Institut Amsterdam, 2000.

[2]   Brundtland Commission, World Commission on Development and Environment: Our common future. Oxford University Press, 1987.

[3]   B. Blättel-Mink, Wirtschaft und Umweltschutz, Grenzen der Integration von Ökonomie und Ökologie, Frankfurt/New York, 2001.

[4]   D. Cansier in O. Renn (Hrsg.): Nachhaltige Entwicklung – Zukunftschancen für Mensch und Umwelt, Berlin u.a., 1996.

[5]   M. M. Cernea, Using Knowledge from Social Sciences in Development Project, World Bank Discussion Paper, Washington, 1991.

[6]   G. Chichilnisky in Social Choice and Welfare, 13, 1996.

[7]   H. Daly in Jahrbuch Ökologie (Ed. G. u.a. Altner), München, 1994, p. 147–161.

[8]   H. D. Feser in Integrierter Umweltschutz – Umwelt und Ressourcenschonung in der Industriegesellschaft (Eds. H.-D. Feser, W. Flieger, M. von Hauff), Regensburg, 1996.

[9]   F. Hasslinger in Neuere Entwicklungen in der Umweltökonomie und -politik (Eds. H.-D. Feser, M. v. Hauff), Regensburg, 1997.

[10]   M. v. Hauff in Zukunftsfähige Wirtschaft – Ökologie- und sozialverträgliche Konzepte (Ed. M. v. Hauff), Regensburg, 1998.

[11]   V. Hauff, Unsere gemeinsame Zukunft – Der Brundtland-Bericht der Weltkommission für Umwelt und Entwicklung (Ed. V. Hauff) Greven, 1989.

[12]   S. P. Huntington, The Clash of Civilisations and Remaking of World Cultures, Simon and Schuster, New York, 1996, D. Landes, Cultural Counts Conference cosponsored by the Government of Italy and the World Bankwith the cooperation of UNESCO, Florence, 2000.

[13]   J. Martinez-Alier, G.Munda, J. O'Neill in The Sustainability of Long-term Growth (Eds. Munasinghe, O. Sunkel, C. de Miguel), Northampton, 2001, pp. 34–56.

[14]   F. C. Matthes in Nachhaltige Wirtschaftsweise und Energieversorgung: Konzepte, Bedingungen, Ansatzpunkte (Ed. H. G. Nutzinger) Marburg, 1995.

[15]   H. G. Nutzinger in Zeit in der Ökonomik – Perspektiven für die Theoriebildung (Eds. B. Biervert, M. Held) Frankfurt, 1995.

[16]   D. W. Pearce, and G. D. Atkinson, Measuring Sustainable Development, Ecodecision, June, 1993.

[17]   D. W. Pearce, and R. K. Turner, The Economics of Natural Ressources and the Environment, London, 1990.

[18]   R. Mohan, M. W. Wagner, J. Garcia, J., Measuring Urban Malnutrition and Poverty: A Case Study of Bogota and Cali, Colombia, World Bank Working Papers, ISSN 0253-2115, Washington D. C., 1981.

[19]   H. Rogall, Neue Umweltökonomie-Ökologische Ökonomie, Opladen, 2002.

[20]   J. Rawls, A Theory of Justice, Cambridge/Mass, 1971.

[21]   G. Seeber, Ökologische Ökonomie – Eine kategorialanalytische Einführung, Wiesbaden, 2001.

[22]   J. Sherman, "Arm's Length", Armed Forces Journal International, 138(2), 2000.

[23]   V. Thierry, No Life Without Roots: Culture and Development, Zed Books, London, 1990.

[24]   UNESCO, Our Creative Diversity, Report of the World Commission on Culture and Development, UNESCO Publishing, Paris, 1995.

[25]   UNDP, "On Overcoming Human Poverty", UNDP Poverty Report, 2000.

[26]   G.Vornholz in Neuere Entwicklungen in der Umweltökonomie und -politik (Eds. H.-D. Feser, M. v. Hauff), Regensburg, 1997.

[27]   World Bank, World Development Report 2002/2001: Attacking Poverty, Oxford University Press, London, 2000.

[28]   World Bank, Voices of the Poor: Can Anyone Hear Us? ", Oxford University Press, London, 2000.

14 Sustainability in Latin American Countries: Challenge and Opportunities for Argentina

Monica Renner

University of Buenos Aires, Ada Elflein 3761 11A (B1637 AMG), Buenos Aires, Argentina, monicarenner@yahoo.com

14.1 Introduction

Research has determined that the economic growth and increase in the Gross National Product of a country are the result of material and non-material factors. The natural resources, human culture, efficiency and implementation of new technologies apart from the capital, are some of the components of the growth. In Argentina as in other Latin American nations, lack of capital and investments requires us to think about other strategies of growth. In the forthcoming years it can be foreseen, there will be shortage of investment, both on a state and a private basis. So there ought to be more stress on non-material factors of growth, such as developing the right vision, leadership and new skills in strategic thinking.

The government will have to make use of the scarce resources in an intelligent way, in particular for its operation and for social aims. Private companies are suffering from lack of credit and unavailability of money to finance working capital. Consequently, this scenario demands that capital be efficiently invested.

The State will have to impose strict criteria and priorities in the performance of public works and the administration will have to carry them out as quickly as possible to create new employment opportunities. Due to the fact that small and medium sized companies do not have enough capital to invest in new machinery and equipment, they will have to be integrated with other enterprises in order to profit from their core competences, apart from having to study the most productive capabilities of their equipments.

It is frequent that industrial owners believe that they are working at full speed when the truth is that they can be much more productive. Nowadays we are essentially governed by the desire to fight poverty and set up lasting economic and social structures.

It is possible to have a better result by focusing only on special areas. Development policies will make a crucial contribution to this end. Concrete measures include the es-

Global Sustainability. Edited by P. A. Wilderer, E. D. Schroeder, H. Kopp
Copyright © 2005 WILEY-VCH Verlag GmbH & Co. KGaA, Weinheim
ISBN: 3-527-31236-6

tablishment of basic elements. Certain activities constitute an enormous strain on the limited financial and expert human resources available.

We have to focus on special areas in which engagement is significant and also promises to be successful. In practice this implies a more specific set of priorities than the ones existing at present. On the one hand, Argentinean projects will be better integrated into the South America Trade Community (Mercosur) in the near future and with European Union markets in an effort to enhance effectiveness. It is important to carry out holistic partnerships to improve results.

As a consequence some regional focal points that demand innovation have to be created. From a realistic point of view today, it must be understood that only through concentration on a few projects will there be change in the long-term process.

14.2   Core Areas for Sustainable Development

Latin America, as an economy in transition, has to create, adapt and innovate. This is the way to develop faster and to find viable solutions, so that we can reclaim optimism. The following aspects play a very important role.

Food industry:
In Argentina, Soya production was increased by a factor of three through application of biotechnology. Cereals and oil products are projected to increase during this year through improved biotechnology and efficiency and only complementarily with investment.

Governments have to support farmers in order to increase sales and create more value for the market. It is important that this support be provided in a manner that causes minimal distortions to free production and trade. In the Uruguay Round, it was agreed that trade-distorting support would be reduced. The objectives of the negotiations include addressing regulations to ensure trade in agricultural biotechnology products based on transparent, predictable and timely processes.

Production and trade of agricultural biotechnology products have increased dramatically in recent years, as new technology has reduced costs, increased yield, and enhanced beneficial characteristics of food and fiber products. This trend will continue in the near future.

It is critical that decision-making be efficient to fulfill the long-run objectives of a fair market-oriented agricultural trading system, as well as helping ensure sufficient agricultural production to meet the world's needs. Extensive trade in agricultural biotechnological products is a relatively new characteristic of internal trade. The transparency in the marketplace is in the interests of producers, who will have clean competition and for consumers, who will have better prices and quality.

Some examples for organizational restructuring and development of European demand-oriented products are as follows:

- Cheese from the south of Buenos Aires Province.
- Honey from Province of Entre Ríos and the south of Córdoba.
- Forestry from the Province of Misiones.

How can Argentina implement new techniques in these core segments? The challenge is the coordination of government, international organizations, and private efforts. It is clear that the industrial sector is responsible for producing goods and services that satisfy human needs.

The National Institute of Agriculture Technology (INTA) [1] has programs for rural development, extension and transfer of agriculture technology to support the integral development of our food-industry. In order to help agriculture and producers INTA has promoted technical progress in the following sectors:

- As regards grains, cereals, sunflowers and oily products supported genetic improvements.
- For the meat sector INTA [1] generated strategic models of control and eradication of illnesses.
- In horticultural products harvest safety programs in the cultivation of potatoes, onions, garlic and legumes to obtain high performance have been developed.
- In fruit, citrus and grapevine production INTA has supported application of integrated pest management.
- The forestry center ensures greater productivity and better quality timber while guaranteeing the sustainability of the natural forests.
- INTA has aided in improvement of the regional production of cotton, wool, peanuts, olive, walnut trees and mate-tea.

INTA [1] opened a skilled Institute of food technology in the western part of the Province of Buenos Aires (Castelar). The protection of arid zones in the Patagonia steep is also a matter of permanent study. Technology programs should be instrumented to provide assistance to economies in transition. This will lead to more economic, environmental and social benefits at national and regional levels.

The generation of appropriate solutions is the result of the associations between INTA and the private sector. Agreements and technology links with companies are important. As a consequence of these relationships a rural change has occurred. We are going from an extensive to an intensive agriculture.

Here is a short description of the general specialties that the INTA [1] helps to address:

- Scenario planning for sustainability.
- Network creation from the beginning to the end.
- Innovative approaches.
- Learning of methodological aspects of foreseeing and back casting.
- Public participation and social inclusion.

These programs address short-term environmental issues and long-term evolution of industrial systems. This possibility requires a number of key conditions to be met to restructure manufacturing and consumers' social effects.

Communication and Information technology (IT):
With the privatization process during the last years, it was possible to modernize and expand communication systems. Today it is feasible to apply new technology to create an environment of speed and flexibility. The fast-changing digital world demands a new approach to the way we structure our business and interact with our customers abroad.
Managers must be ready to embrace new communication technologies and redefine strategies so that organizations are able to adapt themselves to the continuing rate of change.

The development of IT has created ways of reaching global customers. This has to challenge old assumptions, because past business models were based on one-way communication with little feedback about customer satisfaction. Now the global networks with the ability to carry high volumes of data, we have to move the audience in a different way. Fast, flexible, and cost-effective.

We must learn that we cannot stay in the factories waiting for the customers; we have to go out and meet them. As a result of this philosophy we have to participate in international expositions and trade fairs where Argentina may present its products and establish valuable business contacts.

Converting to e-business is a complex process. It requires the conversion of existing processes and IT systems to suit new business strategies. We have to understand the change value of knowledge and learn about the technologies that support e-business.
With the increasing lack of differentiation in quality or price of products and services, the ability to move and share information can be more valuable than the product itself.
Digital technologies provide us with the crucial ability to share real-time information and trade quickly in spite of the fact that we are not situated in the core area of the commercial world.

There is no question that export success is a result of hard work and determination. Nowadays, to be exporters, we have to possess "star" qualities of creativity, initiative and openness to change. The staff has to prove international marketing skills and high levels of customers' service.
These are the questions that I ask to myself when I have to plan an effective strategy:
- Do we work to keep ourselves up-to-date with new technologies?
- Do we have staff with the appropriate awareness of new and future technologies?
- Do we follow the appropriate route to transform todays business into e-business?

This process starts by analyzing existing and future customers' needs and desires. Taking orders on the website, integrating e-commerce applications with back office and adopting an e-business approach is required to put an e-business model in place. We have to use our knowledge of the market to consider how client's expectations will change in the future. In the digital economy, we must be ready for the continual improvement. We have to reexamine our vision; work closely with the staff, suppliers, and partners to implement change.

It is wise to learn to see the organizations from the customer's perspective? The right technology development translates marketplaces into moneymaking business strategies.

Mining:

With deregulation and foreign investments in the mining sector a new horizon has opened. "Bajo de la Alumbrera" [2] is one of the world's major mineral deposits using open pit mining methods. Annual production levels are 120 million tones of total material moved from the open pit.

Through large scale crushing, grinding and flotation processes, Alumbrera's [2] average annual production has increased during the past two years to approximately 200,000 tones of copper metal. The project also produces about 650,000 troy ounces of gold per annum. 6,000 people were required for the construction and commissioning phases. "Mineral Alumbrera" [2] places the highest priority on safety and has developed a workplace safety management program that is focused on all employees.

Minera Alumbrera was the country's first major mining development, and is now a source of ongoing direct and indirect economic benefit and opportunities for communities close to the operations.

Tourism:

Argentina is a big country located in the southern end of South America. Its area covers 3.8 million square kilometers including the Antarctic Sector. A signal near the city of Ushuaia indicates 5,823 kilometers, distance between Fire land (Tierra del Fuego) and La Quiaca. This last city is situated on the northern border with Bolivia. The national territory consists of 23 provinces with a variety of landscapes and climates, which range from subtropical in the North to very cold in Patagonia, but mild climate predominating over the greater part of the country.

There are many natural wonders to offer to the world tourism industry. This kind of service industry has a multiplying effect because a wide range of businesses are involved. As examples, restaurants, hotels, tourist-guides and also souvenirs and leather industry are all supported by tourism. Foreigners coming to Argentina cannot avoid buying handcrafts and regional food products. The benefits of this sector will reactivate our economy in part [3].

Tourists come to Argentina for many different purposes: adventure, pleasure, sports, business, health, exploration, wine tasting and tango dancing.

The government is working to develop Argentina as a tourism recreational destination with high quality and good performance. This is an important means for creating linkages between nature, the needs of human communities and ecological systems. Incorporating opportunities for these linkages includes recreational activities and the maintenance of natural features and wildlife habitat. The Tourism Secretary is a vital part of the tourism planning and the environmentally responsible use of our land. Two major challenges must be addressed. More and more people who demand work opportunities in the tourism sector. Secondly, to maintain and protect the environment, minimizing impacts of increased demands.

For these reasons policies are being developed for promotion of activities and tourism infrastructure. Help is provided to modernize the hotels and small enterprises from the sector and install new renewable energy systems in tourism destinations. The Tourism Secretary inspects the activity in order to maintain quality.

Participation in expositions and tourism affairs has greatly increased in recent years in order to promote this activity worldwide.

The government prepared a wide range of promotion papers, communications and publications to encourage tourism. During the last years farmers discovered a new activity, cattle ranch tourism, to add to their exploitation. La Pampa, the land of the "gaucho" is an immense plain extending 1,200 kilometers from north to south and 600 kilometers in width with the elegance of the "estancias" (ranches) and gauchos in action. It is a real cereal and meat empire. Our cosmopolitan capital by the River Plate, the fascinating city of Buenos Aires, with twelve million inhabitants, is an elegant and immense city with animated nights, theaters and world famous tango.

Forest conservation:

The objectives of the development of different regional projects are to support sustainable production in five pilot areas with fragile natural resources. This localization takes into account the presence of environment degradation, biodiversity, resources conservation and rural poverty.

It is a five-year project that includes a total of 2,500 small producers and their families. In all the cases, forestry is part of the projects. It mainly consists of the settling and handling of the production in agro-forest systems. The extension of project areas ranges from the rainforest to the Patagonia woods [4].

The human factor and the adapted quality of the producer play a very important role in the success of these practices. This is only the beginning of a bigger "Emission Trading Project" that can be used to reduce pollution. In this sense a company can offset its own emissions by causing a reduction of emissions outside its operations because they are buying sources of good air that will be used to compensate for their pollutant emissions.

Regions, States, consumers and businesses can take advantage of this concept by participating in this trade. The emission reduction market gives some companies an added incentive to foster activities aimed at avoiding pollution.

Greenhouse Gas Emission Trade, after the Kyoto Protocol enables to developing countries like Argentina, Ecuador among others in Latin America to sell green gas reduction to others and they have to use them to offset their own commitments, as long as the collective total remains below the agreed levels.

14.3   What are the Consequences?

The creation of efficient working places and the use of new technology requires more trained human resources. This fact will have to be one of the strongest points concerning the growth policy. Otherwise the employees will not be able to be inserted in more sophisticated productive processes.

What I mentioned above demands a change of mentality to be more creative and innovative. This is the only way to sustainable growth for Argentina. Strategic thinking requires the ability to develop teamwork; so as to develop a critical point of view to learn from our own mistakes. We should apply problem-solving tools to help managers confront change. It would be good to plan for transition and envision new possibilities and opportunities.

As we develop a strategic vision, there are different criteria that we should focus on. These will enable us to define ideal outcomes. In this way, we will know the necessary steps to make our developing view become a reality.

In this sense, the following are some criteria to review:

- Organizational structure that involves people and the necessary resources to make it all work. What will the organizations look like? What type of structure will be needed? How can we join people, resources and structures together to achieve the ideal outcome?
- Marketing research. By increasing the power of observation, we will begin to become more aware of what motivates our prospect customers.
- Feedback practices can be used as a tool to help identify critical elements and adjust actions. We have to discover the different points of view from the market, starting from the project planning phase.
- Driving forces that will make our ideal outcome a reality. They might include individual and organizational incentives and values.
- Niche study in the market place that will fit in our business ventures. These core competences, tactics and strategies must be clear.

We should bear in mind that each growing process requires the use of resources and energy, but they are limited. We have to know the existing technology to be applied in a Sustainable Development Scheme for Latin America.

How can we turn Latin America into a more efficient area? International cooperation will help us to go faster and further. We need to take into account the local impacts of the globalization process and adapt ourselves to the external changing world quicker. The increasing demands for goods and services are a result of demographic and economic growth. In which way does this impact us and how can the productive capacity develop in our country to attend this increased demand? We have to analyze which security factors influence savings and investments in Latin America.

The productive structure from Argentina in the last decade remained very recessive. The world economic outlook is that Argentina inserts its products in foreign markets in a competitive way.

What about the labor force and employment? What are the biggest gaps between our values and the employee's behavior in developed countries?

Different people can see things differently and detect potential problems in totally unequal ways. Managing perception differences involves staying alert to the possibility that people might have different perceptions about acting and working. It means becoming aware of something through the senses and the abilities of sighing, hearing, tasting and touching. This is organizational culture. Perceptual differences lead to mutual misunderstandings and communication breakdown will have to be palliated.

14.4 Where Should We Go: Evolution and Tendency?

We have to take an effective approach to create sustainable systems to deliver desired benefit. The inter-relationship with overall business process is currently not well-understood and incorporated into day-to-day management. In our country long term funding of assets is not typically considered as part of the normal business process. Often only short-term political and/or financial needs dominate decisions.

The collection and analysis of data requires improving decision-making processes. The failures are typically the result of a combination of poorly defined scope, failed project management and a lack of business process understanding.

What is the road map?

1) Developing the right vision and strategy
2) Building the data framework
3) Designing the optimized utility
4) Forecasting and planning needs
5) Implementing programs
6) Feedback and continuously improving

Nowadays we are living in a business world; that is increasing connected by networks. Enterprises face transformations; that comprise all the activity chain, from the design and creation of the product and/or services to the product distribution and consumption. We must check all these processes and make use of tools created to help us determine the potential benefits for a better operation.

The training inside and outside the Small Business Enterprises must be constant and in real time. As a first step it is necessary to build a quality infrastructure with the cooperation of the right technologies. We have to satisfy new markets' demands that are willing to buy new quality products at lower costs.

Which aspects are the tendencies, and on which of them should we focus?

- Attract clients' tastes and minds.
- Make alliances with suppliers to minimize products' quality control costs.
- Redefine the distribution mechanisms in an efficient way.
- Personalize markets.
- Increase the speed of launching products to the market.

Global research has revealed tendencies regarding the importance of asset management systems and processes. We have to adapt our strategy and outcomes to this aim.

A number of actions have to be taken in our enterprises before we can create more value. Argentina is a land of great diversity and contrasts. The fact is that the environmental systems are a constant challenge and requires innovative solutions. What are the boundary conditions? This list aims at offering some outlines:

- The application of environmental laws and regulations has to be more efficient.
- Enforcements and penalties in environmental laws are not sufficient.
- Absence of benchmarking practices to improve the application of sustainable methodologies.

A new opportunity is created by which organizations can optimize processes. Today the access to information systems makes it easy to analysis. With the design and implementation of measurements we can develop a business planning system with an accountability framework.

14.5 How Do We Get There: Driving Forces?

The goal is to strengthen the economic performance of enterprises and preserve the environment. The approach consists of methodologies that maximize the use of resources, avoids waste and application of emissions optimizing processes with clean technologies.

The unsatisfied needs are many. The financial resources are insufficient and we have scarce possibilities to obtain foreign capital. At first we have to reinstall the confidence and citizen's dignity by means of solving the urgent problems of our society. And thus we have to encourage hope for a better future. This is a new objective for Argentina and measures will have immediate effects'. Government is trying to satisfy the most elemental needs of the society with the re-establishment of justice, health system and the basic cohabitation rules, so as to offer internal security.

The objectives in the short term are the public sector's reform and rationalization. In the private sector the domestic goods have to be integrated with those from the world economy. There must exist more flexibility and deregulation to have successful access to free market.

The protection of the property rights is very important. Easy access to capital with lower rates and keeping inflation under control are also significant. Which actions should be taken to encourage cultural reforms? Acting clearly, straightforwardly and responsibly will improve the wealth distribution, providing a better education, health, justice and security to the population.

We have to consider the ecological effects of the increased population, longer life expectancy apart from the growing immigration from rural to urban areas to find a better life. This fact makes large cities, like Buenos Aires with 12 million inhabitants collapse.

With the increasing demand for products, needs for more packaging arise simultaneously. The short life cycles of most packaging demands extra attentiveness concerning the selection of materials and the options for reuse that are built into the packaging product. Most regulations worldwide are aimed at reducing the volume and weight of packaging, although this can give a major impulse to manufacturers to choose other options.

By employing Eco-efficient [5] processes companies can diminish the costs of production and site operations. The re-engineering of processes is likely to include a decrease in the use of resources and a reduction of pollution. Some innovative companies not only redesign a product but also find new ways of meeting customers' needs. They work with clients or other stakeholder groups to re-think their markets and re-shape demand and supply. The viability of companies depends on the viability of markets. The globalization of enterprises and markets goes hand in hand with a globalization of social problems and the emergence of global approaches to solving these problems.

Innovative products will succeed on the market when they offer perceptible advantages in use and costs over and above their environmental benefits. For consumers easy and safe handling is often the top priority. For industries improved occupational safety and the avoidance of environmental burdens in production processes is a decided plus. Many customers' needs are satisfied in a material and energy-intensive way.

14.6   Eco-Efficiency Production

Natural resources must be used responsibly. Our society expects that government and private organizations do justice to this task in the long term. The World Business Council for Sustainable Development (WBCSD) [6] defined that Eco-efficiency is obtained by the delivery of competitively priced goods and services. This must satisfy human needs and provide for quality of life. We should reduce ecological impacts and the use of resources to a level, at least, in line with the earth's estimated carrying capacity.

As years went by, this concept developed in different ways, as follows:

- Eco-efficient leadership to improve economic and environmental performance;
- Eco-efficiency as a business link to sustainable development; Eco-efficiency to gain shareholder's international value when global capital will invest in Latin America.

Originally this concept was created by businesses for enterprises, but today it reaches out to governments and intergovernmental organizations all over the world.

During the last years Argentina began to publish reports with cases that cover a wide range of Eco-efficiency practices. The initiative promotes sustainability in enterprises and leads to patterns of savings of money and sustainable behaviors across the regions. Today it is a stakeholder's process to foster joint progress among business, government and society. The implementation of an Eco-label will allow distinguishing environmental friendly products to regular ones.

In a land with so many natural resources, it is very difficult to measure the value of our environment assets, but this is a multidisciplinary task. Some experts create a variety of techniques to quantify the value of these assets. We need to do hard work too, in auditing companies in order to be able to achieve International Standards (ISO).

The way of setting new goals and objectives in the country is:

- Working to re-engineer processes and products.
- Recognizing the priorities to improve recyclable products.
- Intensifying the intelligent uses of energy.
- Being aware of environmental quality, safety and health.
- Providing ISO 14000 standards in industries, to meet international market needs.

It should be wise to promote skills and confidence in order to introduce these actions and thoughts. We have to provide the framework to analyze and apply the concepts of reducing wastewater, air pollution, resource recovery and recycling. Being cautions in recognizing toxic and hazardous waste is an advantage.

This undertaking is based on the concept of sustainable development and responsibility for the protection of natural resources. It means a common pact of security and accountability between the Industry and development policies. Safeguarding the future requires a global course of action.

14.7   Joining Efforts to Sustainable Development

The improvement of safety, health and environmental protection in a company should be a systematic process. It should be transparent and assessable by both internal and external observers. Setting objectives, communicating interim results, and checking that the objectives are achieved all contribute to this. The indicators could be an important management tool The opportunities of making improvements in environment protection and safety can be identified. Ongoing measures can be controlled and progress towards the attainment of objectives might be monitored. Performance indicators are also a suitable aid for communicating the status. Environmental protection policies must become public with reports.

The emphasis in Latin America must be put on products designed to offer greater customers' benefits. For our future global customers we have to develop products and systems tailored to their individual requirements. Some companies belonging to different economy sectors in Argentina, have joined efforts to seek a common goal: sustainable development.

The members from CEADS (Argentine Council of Sustainable Development) [7] have defined an Environmental Policy and Goals, carried out Environmental Audits and put Environmental Improvement Programs in place in order to produce Environmental Impact Studies.

Most companies rank Environmental Management within the 5 top priorities, allocate Environmental costs, they implement Environmental Accounting Systems and require Environmental expertise from suppliers.
Some subjects of study for the enterprises are:

1) Gas emissions:
- Control of fugitive emissions.
- Reduction of greenhouse emissions.
- Eco-efficiency in utilization of gas associated with oil production.

2) Environmental improvements in the design of products:
- Eco-efficiency through reduction of power intensity.
- Replacement by non-toxic materials.
- Replacement of non-renewable natural resources by renewable resources.
- Eco-efficiency in reduction of product packaging.

3) Management Systems:
- Implementation of management systems.
- Integration of environment management – Quality – Health and Safety Systems.
- Integration to the National Quality Award.

4) Waste Management:
- Waste minimisation.
- Recovery and reuse of waste.
- Recycled and recyclable materials.
- Hazard treatment and recovery.

5) Rational use of natural resources:
- Energy efficiency.
- Use of alternative fuels.
- Reduction of waste consumption.
- Water recovery.

6) Corporate social responsibility:
- Development and requirements of environmental quality standards to suppliers.
- Community environmental awards programs.
- Training for safety and environmental teaching staff.
- Foundations that provide educational workshops and postgraduate studies.

The main result that these types of activity give to the enterprises is the generation of corporate confidence and reliability in the business. Communicating leadership is shown through facts. The main role is to offer complete problem solutions to customers, enabling them to focus on core competences. In this case innovation is one key factor to sustainability. The primary purpose is to include more education and training in the organizations.

Shareholders' dialogues and new scenarios with the cooperation of suppliers and partners can provide opportunities to add more value. By following ecological design, it is possible to create products with enhanced functionality. Global demand for some key materials is growing at an unsustainable rate. We have to innovate in order to reduce the material intensity of goods and services.

14.8   Some Cases from Argentina

More enterprises in Argentina have created environment and safety departments reporting directly to the Chairman. They are allocating an increasing quality of resources to there areas, for training and new technologies.

Many companies have already developed new business operations related to environmental services, technologies, recycling, researches and development. The following are three examples of sustainability:

Reuse of Hard sweet residues in ARCOR S. A. I. C. [8]
The production of hard sweets generates around 10% of residues consisting of semi-manufactured; unpacked manufactured and packaged products which, for one reason or another, fail to fulfill the stipulated quality standards. This is added to material obtained

from clean-up of equipment and facilities and products reaching the shelf life date for preferential consumption. A variable fraction of these residues must be managed as waste.

It was decided to used diluted waste streams as food. Therefore residues of hard caramel were also introduced as a source of other processes. The advantages were the re-use of material reducing the sanitary disposal in landfill to zero, with the consequent minimum of the implied environmental and economic cost.

PETROQUIMICA CUYO S. A. I. C. [9] Integration of management systems as a way to eco-efficiency
This company has a 90,000 Ton/year polypropylene production capacity, 50% is oriented to export and the other 50% is sold on the domestic markets. The environmental project includes the implementation and evaluation of the advantages of integrating standardized quality systems with the environmental ISO 14000 standards. This integration brought many advantages without the increment of costs. It was possible to use the same structure and human resources for auditing procedures, training and to do risk analysis.

Quality and Environmental Strategy agreed with EDENOR's suppliers [10]
Edenor started working in a concession area after the intense reform program that has taken place in Argentina. They provide services to 2 million clients in the northern part of Buenos Aires.

Currently the enterprise is developing activities addressed to the assurance of quality and to the environment management, based on ISO-9000 and 14000 standards respectively. The assurance of the quality and control chain was interrupted when the Company suppliers were not integrated to ISO standards. For that reason, the project target was to develop a strategy regarding Quality in the contractors and suppliers from the electric sector and in its auxiliary industry.

14.9   Conclusions

- Governments and Companies must enter into a continuous and open dialogue with the community about past achievements and future priorities on their road to seek peace and bring sustainability. This is the only way to harmonize the needs of business and society in the long term. In each case, a solution must be sought. Our President, Mr. Nestor Kirchner's affirmation is "he will build the necessary infrastructure to profile Argentina as a productive country in the agro-food industry, tourism, energy, mining, transportation and new technologies". This effort will create genuine employment, and requires to join work from the public and private sector.
- One of the priorities is stressed in tourism. This is a good area to reactivate the economy and push the related services industry. We must implement practical solutions. This will result from the interaction of the political framework, and social needs and market opportunities.

- As a developing country, we must compete with quality and good prices in the marketplaces with the products of developed countries. We are working to ward producing goods that meet international standards. In this way, companies can create more workplaces for citizens and a better situation for society.
- Looking for future business viability makes us think that enhancing Information Technology (IT) competitiveness will be a very good opportunity for Latin America, because the speed of IT products development depends on the ability to gain quick and reliable insight into market movements to understand new needs. The world will require more IT in real time and at low costs. Our advantage consists in that we are situated at the same hemisphere with the same time zone as developed states. This will be a good opportunity in comparison with other IT solution suppliers such as India, Ireland and Singapore that are in a different time zone.
- A United Nations Program for developing countries shows that the developed world invests 2.3% from its GNP (Gross National Product) in research and developing of new technologies. In Asia the investments are 1% and for Latin America counties only the 0.6%. With this poor background, governmental authorities should create a massive training program so that a great range of population is involved in acquiring skills for adaptation to new IT technologies.
- In the first four months of this year, the number of workplaces increased 2.8%, which is equivalent to 120,000 new jobs. The increase was focused on four core areas. Building sector 14.2%, Agriculture 7.4%, Mines exploitation 5.7% and Financial Services 2.8%. This acceleration in the economic activity will have positive effects.
- In business organizations effective and direct communication skills are essential. The exchange of knowledge and ideas on environmental and management issues is a source of inspiration to everyone.

Cooperation with other countries would not only attract new trade opportunities, but also resolve environmental problems worldwide. Frequent and friendly communication refreshes our thinking on possible successful solutions.

References

[1]   M. Ruiz, M. A. Perez, Arguello, The National Institute of Agriculture Technology (INTA), Argentina, 2003, 32(2), pp. 34–38.

[2]   Australian Trade Commission, COALAR Council on Australian Latin America Relations, publication Nr. 1, Minera Alumbrera Brochure, 2002.

[3]   M. A. Vicente, Desarrollo Turístico en Argentina, Publisher Ediciones Grupo Macchi, Buenos Aires-Colombia, 2000, pp. 102–125.

[4]   M. Carlos, Publisher Universidad Nacional de la Patagonia San JuanBosco, Instituto Forestal Esquel, Research, Argentina (complete publication), 2001.

[5]   H. Fischer, Umweltkostenmanagement, Environmental Cost Management, Publisher Hansen, Germany, 1996, pp. 1–27.

[6]   Ing. L. Trama, IRAM/ISO Instituto Argentino de Normalizacion publication, Actualidad International article, TC 176 "Novedades en gestion y aseguramiento de la Calidad" Buenos Aires, 1999, pp. 9–21.

[7]   Ing. Florin, CEADS (Argentine Council of Sustainable Development) "Responsabili-dad social empresaria, tendencies globales y regionales, perspectives en la Argentina", publication by CEADS, Latin America, Chapter 3–6, 2003.

[8]   Ing. M. Carranza, Eco – Eficiencia Arcor Article "La Providencia, Mecanizacion de cosecha de cana y reuso de residuos de la produccion de azucar" Buenos Aires, 2003, 14–17.

[9]   Felix de Bulhoes, Relatorio of Sustainable Development Publishing in Brasil, 2002, pp. 178–190.

[10]   S. Wiel and J. E. McMahon, Lawrence, Berkeley National Laboratory USA Energy-Efficiency Labels and Standards, a Guidebook for Appliances, Equipment And Light-ing, Argentine Council of Sustainable Development, Chapter 2, 2001.

15 On the Roles Engineers May Play in the Attempt to Meet Basic Demands of Man and Nature

Peter A. Wilderer*, Mirka C. Wilderer**

* Institute of Advanced Studies on Sustainability, European Academy of Sciences and Arts, c/o Technical University Munich, Lehrstuhl für Wassergütewirtschaft, Arcisstraße 21, 80333 München, Germany, peter@wilderer.de
** Moosweg 5, 83727 Schliersee, Germany, mirka@wilderer.de

15.1 Millennium Development Goals

On September 8, 2000 the General Assembly of the United Nations resolved the UN resolution 55/2, called "United Nations Millennium Declaration" [1]. By signing this document all 191 member states have pledged to pursue to following eight goals:

1 Eradicate extreme poverty and hunger.
2 Achieve universal primary education.
3 Promote gender equality and empower women.
4 Reduce child mortality.
5 Improve maternal health.
6 Combat HIV/AIDS, malaria and other diseases.
7 Ensure environmental sustainability.
8 Develop a global partnership for development.

At first glance these eight goals become very ambitious "engineering" issues when reading the detailed descriptions and the deadlines set to reach the specific goals. In particular, it was resolved to halve by the year 2015 the portion of people who suffer from hunger and who do not yet have sustainable access to safe drinking water and sanitation. Additionally, the member states promised to integrate the principles of sustainability into their policies and programmes as well as to reverse loss of environmental resources.

Obviously, there is innovative and highly integrated technology needed to be able to meet these goals. Engineers from all over the world are called upon to develop methods capable to reduce hunger and thirst, to provide sanitation, and to lower the demand of

Global Sustainability. Edited by P. A. Wilderer, E. D. Schroeder, H. Kopp
Copyright © 2005 WILEY-VCH Verlag GmbH & Co. KGaA, Weinheim
ISBN: 3-527-31236-6

natural resources and energy – all this within a very limited period of time and for a price affordable by the worlds economy.

It is probably not so important to argue whether these goals will be actually met by the year 2015 and whether the required technical means will be available in time. Tremendously important is, however, that the message has been received by State authorities, goals have been agreed upon and action has been promised to be taken. To underline the need for action the General Assembly decided in the course of its session on December 11, 2003 to declare the period between the year 2005 and 2015 the "The Water Action Decade: Water for Life".

In the catalogue of goals poverty, hunger, lack of safe drinking water, lack of sanitation, and participation in the process of decision making are key words which are particularly important with respect to the discussion on sustainable development. The significance of these keywords becomes striking when one considers that the fraction of the world's population undersupplied with the most elementary resources of survival is tremendously high. Over 1.2 billion people are estimated to be without access to safe drinking water, for instance, and twice as many are estimated to live without access to any reasonable sanitation.

It is, however, not the absolute number of people lacking water and sanitation that should make us alert. Even more critical is that the fraction of poor and undersupplied people is sky rocking in many urban and peri-urban areas. These people who are hungry, thirsty and threatened by water born diseases are concerned mainly about surviving the next day and it is hardly possible to motivate them to think about and actually contribute to any kind of sustainable development of the local community and the country. Therefore, to be able to make progress in sustainable development and to obtain conditions under which future generations can enjoy prosperity a very first step has to be taken: reduction of the number of people living under critical conditions. Meeting the Millennium Development Goals has to be understood as a key to sustainable development.

15.2   Engineers and Sustainability

The term "ecology" is commonly defined as "the science of the relationship between organisms and their environment". The term "environment" is understood as "the integrated effect of all factors (environmental factors) affecting growth and proliferation of organisms living in a certain biotope". In this context, biotope stands for a region inhabited by populations of microorganisms, animals and plants, but also by people, industries and governmental institutions.

It is worth realizing that organisms are components of an environmental system but act as environmental factors at the same time. The same is true for engineers. Through their actions engineers are major modifiers of the environmental situation in a biotope. Engineers, directly or indirectly, change the environment in the attempt to fulfill various needs of the human society. From looking at Figure 15.1 the question arises: Who defines what the human society really needs?

Figure 15.1 Flyer of a telephone company operating in Kenya asking people to connect to the countries mobile network. In the same countries numerous people have no access to safe drinking water and sanitation: A contrast exemplifying the power of commercial advertisement

The term "sustainability" describes a situation characterized by a certain quality remaining dominant for a foreseeable period of time – the quality of agricultural land, for instance, as a source of foodstuff, a forest as source of wood (see Chapter 1) or the quality of a lake as source of water for drinking, for irrigation purposes, or for producing goods such as paper or steal. Sustainable development is a process leading ultimately to sustainability.

Throughout modern history, engineers have championed to enhance quality rather than maintain it. A slogan often used in the engineering society is: the "better" beats the "good". The question here is: Who defines what is good or better?

Engineers are problem-solvers. For instance, a river might be seen as a problem because it hinders easy travel from one side to the other. To solve this problem, engineers build bridges. Engineers are good in building bridges also in the metaphorical sense. However, what purpose the "bridge" is eventually used for is a different story. It can be used for good or bad as the September-11 events have shown.

The potential of abusing technology is, however, no reason to demonize it. Nevertheless, the question remains whether technology is the right means to solve the problem of maintaining quality, thus of guaranteeing sustainable development? The "Technikfeindlichkeit" (aversion to technology) which has developed over the past years – in different cultural regions with different intensity – speaks a different language.

Engineers act on demand of the society, but they also influence the evolution of novel demands by creating technical innovations. The reciprocal influence of vital demands and technological innovations has been a major driving force of the unique economic development in the industrialized countries over the past centuries (see Chapter 3). It may also drive the development towards sustainability, but only if the society addresses a strong and enduring will to accomplish sustainable development. For instance, if the society would value long lasting products higher than fashionable items the engineer would find ways to design and produce those products. But is our society, both in the developed and in the developing word, mature enough to give up on fashion?

Fashion is often seen as means to enhance quality of life. People feel good, for instance, wearing up-to-date clothes. In this respect, fashion is also a significant driving force of economic growth. Commercial advertisement awakens demands even when real needs do actually not exist (see Figure 15.1). On the other hand, new demands provoke technological progress. And progress of technology facilitates life, as the invention of the wheel clearly has demonstrated.

This leads to the assumption that technology and the progress of technology are important factors in the attempt to secure conditions under which mankind can sustain on this planet. A closer look, however, reveals that progress and application of technology is just one factor among many others. As Ernst Ulrich von Weizaecker points out in his book "Factor 4" efficiency is another important but often underestimated factor [2]. He assumes that halving the use of "nature" but doubling efficiency would enhance our efforts to achieve sustainable development by a factor of 4.

15.3   Design of Technology in a Globalizing World

The common argument goes that in the course of globalisation technology that has been developed in the industrialized countries has become known and available throughout the world (see Figure 15.1) causing a certain degree of cultural convergence. Does this mean that regional differences in quality demands will become extinct within the very near future? Will our world gradually transform into uniformity, not only technologically but also with respect to economic and societal conditions and customs? And is this desirable at all?

The ecological consequences of such a transformation would probably be very similar to those in agriculture where large-scale mono-culture systems have been predominantly used in many parts of the world for decades. The experience gained in agriculture may be worth considering while assessing the consequences of such a trend towards homogenisation and uniformity.

Giving up on bio-diversity means giving up on self regulating processes within an eco-system. As a consequence, regulation must be accomplished by man. And this may well lead to a catastrophe due to the evolution of non-intended side effects. Long term effects have not been properly respected in the past, be it because they have simply not been known or because they are neglected (see Chapter 7). As a conclusion, we must consider bio-diversity – and in analogy to this but on a different scale: cultural diversity – as important factors in respect of sustainability.

15.3.1 The Cultural Dimension

Culture has been defined in various ways. According to Kluckholn [3] culture is "patterned ways of thinking, feeling and reacting, acquired and transmitted mainly by symbols, constituting the distinctive achievements of human groups, including their embodiment in artifacts." Common history, family and tribal traditions, languages, local dialects, poetry, fairy tales, and religious beliefs: all belong into the category of cultural value influencing human action. Culture is hierarchically structured and connected reaching from specific codes of conduct to common shared values. In this respect, Schein [4] distinguishes the following levels of culture (Figure 15.2):

- The level of artifacts which are visible, can be directly observed, but are difficult to interpret, such as language, symbols, legends, art, or technology.
- The espoused values which are partially visible and partially unconscious such as standards, commands, and prohibitions as well as strategies, goals and philosophies.
- The level of basic underlying assumptions and convictions which are invisible and unconscious. This level represents taken-for-granted beliefs, perceptions, thoughts, and feelings such as the image of man, the relationship of man and nature, or the approach towards technological change.

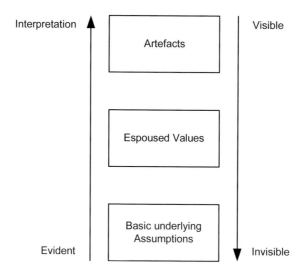

Figure 15.2 Levels of culture according to Schein [4]

Cultural diversity, in this context, means the coexistence of different standards of value and quality measures within a certain region. Cultural diversity includes differences as to how traditional and metaphysical values are acknowledged, the value of materials in general, and particularly of natural resources like air, water and soil, but also the value of technological advances.

From a dynamic perspective, however, culture is not static, but evolves over time. Hofstede [5] describes culture as "collective programming of the mind which distinguishes the members of one group from another." Nevertheless, humans are never fully determined by their cultural roots. Instead they always have the possibility to diverge from these given patterns in their specific action and behaviour. Tung [6] therefore defines culture as "an evolving set of shared beliefs, values, attitudes, and logical processes which provides cognitive maps for people within a given societal group to perceive, think, reason, act, react and interact." This conceptualization of culture seeks a middle-ground between those who support cultural divergence (cultural imperative), on the one hand, and those who insist upon cultural convergence (technological imperative), on the other hand.

Representatives of the cultural divergence approach argue that significant cultural differences will remain across people of different societal groups over time. In opposition to that, the technological imperative school of thought asserts that as countries around the world embrace universal technology, cultural differences will no longer matter. In this respect, one may argue that as a result of the modernization and globalization fundamental structural changes and societal transformations will proceed towards an information society [7], a post-industrial society [8], or a knowledge society [9, 10]. Common to all of these concepts is the idea that the production, reproduction and distribution of information and knowledge is a constitutive principle of modern societies. This also implies a gradually decreasing material demand. As a result, pressure would be taken away from the need of natural resources and facilitating sustainable development. Assessments like this, however, over-emphasise the power of technology by far.

While it is undoubtly true that globalization and technological and economic development among other aspects have served to bring cultures around the world closer together than ever before in their evolutionary processes, there are some cultural aspects, such as deep-seated values, traditions, and ways in which information is processed and interpreted, which are deeply rooted and thus not readily changed in the very short future.

Is it really desirable that cultures converge due to globalization processes? Even when mobile phones and e-mail communication are available in the rural areas of Kenya (see Figure 15.1), when a personal tragedy, such as a severe accident, illness, or death of a beloved relative occurs the importance of technology diminishes and other factors become more crucial. People need help during such a crisis not only technologically but psychologically as well. All at once traditions gain importance, the long neglected prayer, songs learned during childhood, the hand of a close friend or a family member caressing ones cheek.

15.3.2 The Global Dimension

Much of the talk of "globalization" is confused and confusing. "Globalization" has become a buzzword, but those using the term often have contrasting understanding. There are few terms which are in fact as poorly conceptualized as globalization. But what does it really mean? Scholte [11] identifies at least five key definitions of globalization that are in common usage. They are related and overlap, but the elements they highlight are significantly different. In this respect, globalization is commonly used in the sense of internationalization, liberalization, universalization, westernization or modernization, or deterritorialisation. Of these five approaches, it is only the last, according to Scholte and many others, that offers the possibility of a clear and specific definition of globalization. Among this line of though, Held et al. [12] define globalization as a "process (or set of processes) which embodies a transformation in the spatial organisation of social relations and transactions – assessed in terms of their extensity, intensity, velocity and impact – generating transcontinental or inter-regional flows and networks of activity." They specifically point to four factors of globalization, namely its a) geographical extension, b) condensation of the good and capital markets, c) implications on the activities and power of national and local actors, and d) networks and infrastructure supporting intercontinental transactions.

Held et al. [12] divide the controversy over globalisation into two opposed positions: the "hyper-globalizers", on the one hand, and the "globalization sceptics", on the other hand. The hyper-globalizers tend to be linked ideologically to business. The best example of them is the Japanese business writer Kenichi Ohmae [13]. According to his view, globalization is changing everything. Globalization means the expansion and intensification of the global marketplace. The dominant feature of our time is the intensified economic interdependence and competition among all economies. This process may go so far that nation states loose most of the power they used to have.

A direct opposite point of view is taken by the "globalization sceptics" who argue that "there is nothing new under the sun". They are sceptical of the idea that there are big changes going on in the world and that the world is becoming more integrated than it used to be. This approach is represented by Hirst and Thompson [14], who argue that if one looks at statistics on global trade, globalization was more highly developed at the turn of the century than it is now.

Seeking a middle ground between these two opposing position, the most prominent concept on globalisation is the work of Anthony Giddens [15]. He argues that both the views of the hyper-globalizers and the globalization sceptics are wrong. According to him and contrary to the hyper-globalizers, the world is currently at the beginning, not at the end of the process of globalization – a fundamental shake-out of world society, which comes from numerous sources. Contrary to the globalization sceptics, he argues that globalization is the most fundamental set of changes going on in the world today. It is not advanced as far as the hyper-globalizers say, but it is still the most fundamental phenomenon of our times. Additionally, globalization should not be understood as a wholly economic concept and purely driven by economic market imperatives. The economic marketplace is certainly one of the driving agencies of the intensified globalization. However, globalization refers to a set of changes, many of which are of social, cultural and political nature. It is a multidimensional construct which embraces all societal areas and fields of action, such as the economy, politics, crime, media, and culture. Globalization therefore does not have a single force pulling in one direction, but is a complex set of partially contradictory forces.

For Giddens [15] globalization is a contradictory and uneven process. He uses the idea of dialectics to express a certain dynamism and defines globalization as "the intensification of worldwide social relations which link distant localities in such a way that local happenings are shaped by events occurring many miles away and vice versa." This refers to the idea of a combination of absence and presence: social relations across space, time and social context. Giddens points at the phenomenon of "action at distance". According to him, globalisation refers to the increasing inter-penetration between individual life and global futures. A larger and larger number of people live in circumstances in which globalized social relations organise major aspects of day-to-day life. In term of the individuals, Giddens [16] notes that "globalization is not just 'out there' – to do with very large-scale influences. It is also an 'in here' phenomenon, directly bound up with the circumstances of local life." Giddens therefore refers to globalization as a process which simultaneously "pulls away from" local communities and nation-states, yet also "pushes down" on those same communities and nation-states. Local communities' beliefs and cultural values may be "pulled away" and globalized in the sense that these once particular and local beliefs and cultural practices are universalised across the globe. Individuals and social groups may experience this universalization as a "dilution" and "corruption" of their cultural beliefs and practices and may seek to resist the process, sometimes with violence. Fundamentalistic Islamism is an example for this. At the same time, globalization "pushes down" and presents local communities with new possibilities and demands.

It is this dynamism that Giddens captures in the dialectics of globalization and localization, of opening and closing, of dis-embedding and re-embedding. Localities remain existent for social relations, however, now with a global reference. People always act locally, but now with a global meaning. We are currently able to observe a globalization of the social practices of the locals as well as a re-embedding of globalization in the local context – in the practices of the locals. Globalization and localization (Figure 15.3) are therefore two sides of the same coin. Disembedding processes can only be sustainable, if re-embedding processes are guaranteed.

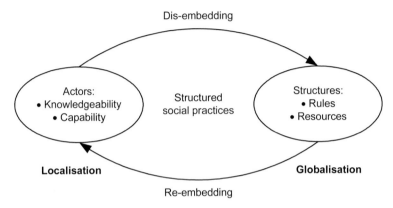

Figure 15.3 Giddens [16] notion of "globalisation" and "localisation"

15.3.3 Synthesis

Advanced global technology is to be understood as a blessing, but equally are local traditions which enable a person, a family, a tribe, a country to cope with often rapidly changing economic, political or climatic conditions. Does a society which has lost ties to its indigenous traditions have a chance to sustain? Technology can help in achieving sustainability but in addition we also have to make sure that the adaptive capacity of the various societies in the world is maintained. Only when the re-embedding of the globalization in the local context is guaranteed, can sustainability be achieved. Technology can solve problems only when tailored to the specific needs of specific applicants living in specific areas of the world. The engineer who must develop such tailored solutions must respect not only physical laws but the whole range of non-physical boundary conditions governing the very local situation. Following Talcott Parsons [17] theory of structural functionalism according to which social systems consist of different subsystems all of which serve different functions in order to maintain the overall system, solutions need not only to concentrate on technical and economic aspects, but also have to take the political, social and cultural subsystems into consideration in order to present sustainable solutions (Table 15.1).

Table 15.1 Subsystems of society and its functions [17]

Political System	Economic and Technological System
Goal Attainment	Adaptation
Social System	**Cultural System**
Integration	Latent Pattern Maintenance

In sum, development and implementation of technological solutions must correspond to the very economic and technological, political, social and cultural subsystem of the people living in a specific region. Two approaches lend itself in order to bridge current gaps:

- On the one hand, the level of education of the local community needs to be raised fundamentally. People should reach an educational status which allows them to express their expectation and views in a "language" comprehensible by engineers. This seems especially relevant in respect to the subsystem "economy and technology" serving to maintain the adaptation of the overall system.

- In turn, engineers must be provided with methods enabling them to gain knowledge about the specific political, social and cultural subsystems of the local community. Concerning the cultural system providing pattern maintenance, engineers need to be able to understand the regional needs and fears, values and norms, priorities and concerns of the people stemming from the local traditions and religious beliefs in order to be able to move the technological development into the right direction.

Only when both sides are equally respected, only when both disembedding and re-embedding mechanisms are ensured, can sustainable technology be developed and implemented. Only then can we move from one-sided communication in the sense of engineers informing on technological possibilities to two-sided communication commonly known as dialogue.

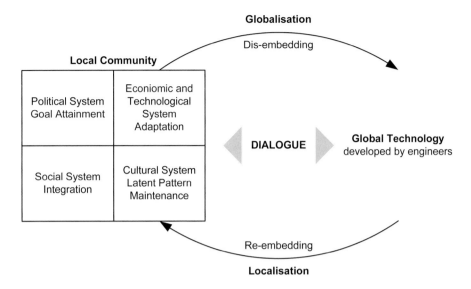

Figure 15.4 Bridges to be build

Only when these bridges are built (Figure 15.4) can irrational resistance against technological solutions and outbreak of "Technikfeindlichkeit" be avoided. Only then can an
ecologically, economically, social and cultural sound development towards sustainability
be expected and a more peaceful coexistence of the countries on earth can be achieved.

15.4   Urgent Questions to be Answered

The questions raised above provoke a great number of questions when trying to define
the role engineers may or should play in the attempt to transfer the sustainability axiom
into concrete actions. For instance:

- What relations do people have to natural resources such as water, air, soil, mineral
  deposits, people of different economic status and political, social and cultural background?
- How can we put the idea of social and cultural sustainability in practice? How can we
  achieve re-embedding processes in reality? How can we realize that people are involved and participate in the development and implementation of technology in order
  to make it socially and culturally sustainable? In simple words, how can we guarantee the social and cultural sustainability of technological solutions?
- Can sentiments, emotions and traditions ever over-rule commercial interests?
- How do we curb profit-driven commercialism in the exploitation of natural resources?
- What position do the world religions hold with respect to the sustainability issue?

- Can moral high ground in sustainability ever be established and maintained when commercial exploitation of natural resources is understood as a prerequisite of better life?
- The root cause of excessive exploitation of resources is the uncontrolled population growth. Which position do the world's religions take in this respect?
- What do we need to accomplish to make ecologically sound technology acceptable?
- Which technology is ecologically sound, and which is not?
- How are advances in technology valued by people of different cultural heritage?
- Is education a means to influence consumer habits, and if so, who is responsible for the educational curricula?

15.5   Conclusions

1. Meeting the UN Millennium Development Goals is to be understood as a primary and very crucial prerequisite in the attempt of getting sustainability measures implemented in our societies, economies and governance, locally and worldwide.
2. Engineers are called upon to develop methods capable of making the UN Millennium Development Goals effective.
3. It must be understood that technology by itself cannot solve the problems addressed in the Millennium Development Goals. Required are multi-disciplinary and multi-cultural efforts.
4. Technology can solve problems but only when tailored to the specific needs of the customers. The engineer who is supposed to come up with tailored solutions must respect not only physical laws but the whole range of non-physical boundary conditions governing the very local situation.
5. Engineers must gain knowledge about political, social and cultural situation of the specific region or local community. Simultaneously, people must become educated so that communication between engineers and applicants of technology can move from one-sided informing to a two-sided dialogue. Only then can the outbreak of Technikfeindlichkeit be avoided and irrational arguments overcome.
6. Technology can help in achieving sustainability but in addition we also have to make sure that the re-embedding processes of globalization are realized. A fine tuned balance between pride of their cultural heritage and readiness to take advantage of the "new" must be achieved.

References

[1]   United Nations Millennium Declaration, can be found under
        www.un.org/millennium/declaration/areas552e.htm, 2000.

[2]   Ernst Ulrich von Weizaecker, Faktor 4, Doppelter Wohlstand – halbierter Naturver-
        brauch. Droemer Knaur Verlag, Muenchen, 1997.

[3]   C. Kluckholn in The Policies Sciences (Eds. D. Lerner, H. D. Lasswell), Stanford
        University Press, Stanford, CA, 1951.

[4]   E. Schein in Organizational Culture and Leadership, Jossey-Bess Publ., San Francisco,
        CA, 1996.

[5]   G. Hofstede in Culture's Consequences, Sage Publ., Beverly Hills, CA, 1984.

[6]   R. L. Tung in Virtual O. B. Electronic Data Base (Ed. F. Luthans), New York:
        McGraw-Hill Inc, 1995.

[7]   F. Machlup in The Production and Distribution of Knowledge in the United States,
        Princeton: University Press, Princeton, NJ, 1962.

[8]   D. Bell in Die nachindustrielle Gesellschaft, Campus Verlag, Frankfurt a.M, 1985.

[9]   P. F Drucker, The Atlantic Monthly 1994, 273(11), Boston, MA.

[10]   H. Willke, Systematisches Wissensmanagement, UTB/Lucius & Lucius, Stuttgart,
        1998.

[11]   J. A. Scholte in Globalisation, A critical introduction, Polgrave Publ., London, 2000.

[12]   D. Held, A. McGrew, D. Goldblatt, J. Perraton in Global Transformations – politics,
        economics and culture. Polity Press, Cambridge, USA, 1999.

[13]   K. Ohmae in The Borderless World, Power and Strategy in the Interlinked Economy.
        Harper, New York, NY, 1990.

[14]   P. Hirst, and G. Thompson in Politik der Globalisierung (Ed. U. Beck), Suhrkamp,
        Frankfurt/M, 1996.

[15]   A. Giddens in The Consequences of Modernity, Stanford University Press, Stanford,
        CA, 1990.

[16]   A. Giddens in Beyond Left & Right: The Future of Radical Politics, Polity Press,
        Cambridge, USA, 1994.

[17]   T. Parsons in The Social System, Free Press, Glencoe, 1951.

16   Integrating Cultural Aspects in the Implementation of Large Water Projects

Martin Grambow

Bavarian State Ministry of Environment, Health and Consumer Protection, Rosenkavalierplatz 2, 81925 Munich, Germany, martin.grambow@stmugv.bayern.de

When considering sustainable development, Agenda 21 is still the guideline for necessary global action.

Based on Agenda 21 the concept of Integrated Water Resources Management (IWRM) was developed to address the particularly critical water area. The IWRM sets the norms for an integrated approach that for quite a long time experts have considered to be the only suitable water management method.

Basically, therefore, it appeared as if the solution had been found. In practice, however, the actual problem today lies in how to get *"from vision to action"* based on this approach (John Brisco, Senior Water Advisor in The World Bank, Closing Plenary of the water week Washington March 2003). There obviously seems to be issues in the apparently comprehensive IWRM approach that need further investigation. Project designers nowadays tend to focus mainly on the technical and organizational project aspects. Is this enough to achieve the explicit aim of sustainability through integrated approaches?

Some practical examples related to the activities of the Water Technology Transfer (*Technologietransfer Wasser* – TTW [1]) Project seem to prove the need of integrating cultural aspects into projects. Cultural influences reveal themselves at two equally important levels: "implicit" working/operating culture, which influences the internal project implementation, and "explicit" culture as project goal and purpose as a need and a part of human quality of life.

These approaches provide exciting insights into the parameters for success in international water projects. They raise the question of to what extent the "missing link" in the water infrastructure approaches is to be found at the cultural, spiritual level.

Global Sustainability. Edited by P. A. Wilderer, E. D. Schroeder, H. Kopp
Copyright © 2005 WILEY-VCH Verlag GmbH & Co. KGaA, Weinheim
ISBN: 3-527-31236-6

16.1 General Conditions

Something needs to happen! The larger picture is well known: billions of people are without access to safe drinking water, without proper sewerage, children and adults are in pain. Suffering and awareness as to the need for action are growing. With the Millennium Development Goals (the MDG, which envisage halving by 2015 the number of people who do not have access to safe drinking water and adequate sewerage systems) the UN has taken this task upon itself – and along with it certainly a great challenge!

While the present paper addresses the issue of improving project implementation by integrating cultural aspects into practical work, it should not be concealed that apart from this there are basic questions, which are still open.

Beginning with the use of resources and up to social challenges, global systems obviously lack stability (and will continue so for quite a while). In spite of all their capacity the self-regulating forces of the world market alone seem unable to create the desired stability. Among experts there is no further doubt that one of the main tasks is a better internalization of environmental and social (including cultural) costs and values as proposed by the Club of Rome [2].

Regardless of the question of whether permanent growth is the prerequisite for prosperity, growth itself is already determined by the demographic evolution (see Chapter 12). At present, however, all growth causes an even greater consumption of resources, including water. The minimum goal should be avoiding wasting of resources. Improved ecological efficiency is a step in this direction. A further step could be a fairer distribution of profits to be achieved from future growth as, suggested by Weizsäcker, i.e. the so-called Factor 10 Club. The Factor 10 concept draws on a tenfold increase of ecoefficiency over the next hundred years, i.e. with the same quantity of resources used today, particularly energy, ten times as many goods could be produced. Rademacher further elaborates this theory distributing growth in a ratio of 4 and 34 (4-fold growth for the industrialized countries, 34-fold growth for the developing countries). The result is the 10 : 4 : 34 approach [3].

That means, one must look about for more efficiency and what is best to do with the "saved" resources (values). Seen optimisticly, improved efficiency gives the answer to the question of financing.

Regardless of which path turns out to be more viable in detail, the general conditions under which sustainable solutions can emerge have to be examined. F. J. Rademacher makes the following observation in 'Balance or Destruction' (*Balance oder Zerstörung*): "*The key issue is how to organize economies. This is not a question related to business administration but to national economy. The answer to this question is both trivial and extremely difficult. We generally only address the trivial aspect: Economy is based on competition. Everybody understands this at once. Much more crucial, however, is the second aspect, i.e. the general conditions under which competition takes place. This involves primarily public interests, social issues, conservation of cultural diversity and protection of the environment. The general conditions determine what cannot be changed and is to be maintained with regard to social facts, the diversity of cultures and intactness of the environment.*" The definition of the general conditions always has considerable influence on the prospects for project success.

An important prerequisite for progress is a constantly increasing knowledge about the concept of sustainability (for some problems see chapter 8). A very comprehensive definition of sustainability specifically related to the water area is given by Kraemer with the 9 sustainability criteria for water management [4].

In recent years, much has been going on in respect to the development and implementation of appropriate systems. The activities of the UN and the highly participatory expert meetings at the large conferences have contributed to a greater consensus on strategic and tactical measures. Thus the World Bank strategies show increasing proximity to the demands of Agenda 21 and of the NGOs.

The intensity with which the World Bank is looking for more integral solutions in the face of the challenge posed by the MDGs also shows in project priorities established since 2003: Under the title "culture and poverty" the influence of cultural aspects on the task of the World Bank – combating poverty – is to be analyzed [5].

So, it is directly or indirectly understood that there is an influence of culture on the core tasks of development banks. However, this influence is normally not defined and is used synonymously with a widely differing range of meaning as to the term and its signification. While social and environmental issues are being quantitatively implemented as a matter of course – at least in open-minded projects – quite often cultural aspects are not even qualitatively addressed.

Within the scope of the Water Technology Transfer (TTW) Project of the Bavarian State Ministry for Environment, Health and Consumer Protection there is the attempt to process these insights concerning the relationship between culture and sustainable development on a practical level, as well as to integrate them into ongoing project work. The base herefore is the interpretation of Agenda 21 as a management model.

16.2   Agenda 21 and Integrated Water Resources Management as a Steering Model

Agenda 21 is nowadays assumed to be widely known, although in practice only a very small number of experts can say of themselves that they have read or even analyzed the paper. E.g. the BFZ (training centers of Bavarian Employers' Associations) organizes together with TTW so called AMBITEC courses: International groups of experts visit Germany, esspecially those working with environmental technology, and discuss in this frame the Agenda 21 approach. One question that always arises is about the level of knowledge of Agenda 21. The same is done with national (German) groups of experts. The result is nearly always the same: All in all the knowledge about Agenda 21 and the involved power seems to be still poor. It is seen as a static paper of arguments, often mentally divided in sectoral approaches.

Thus the value Agenda 21 is quite underestimated! According to the interpretation of TTW, Agenda 21 is seen as a strong steering model (see Figure 16.1).

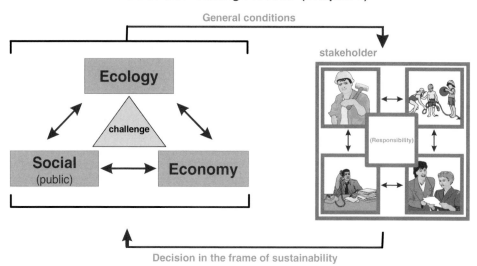

Figure 16.1 Agenda 21 as a management model

The main issues relate to the well-known Agenda 21 triangle formed by Ecology, Economy and Social Factors. Using this interpretation, every serious problem is based on these three pillars. Any analysis should consider the problem dimensions in these three areas. The result will determine the general conditions. In line with the principle of participation, the necessary stakeholder groups are also included in addition to the responsible public and private bodies in order to solve the problem. This does not mean that the responsible bodies are released from their responsibility but that they are able to improve the accuracy of the analysis thanks to participation and through discussion of possible solutions with those involved. At the same time the participatory process provides the additional input of stakeholder knowledge. The whole group – consisting of the responsible bodies and the stakeholders – is in principle completely free as to its decision with the only restriction being that the solution found has to be sustainable. This, of course, is the actual challenge. At the same time it is stated that every sustainable solution in its turn almost necessarily comprises the three areas – social factors, economy, ecology. The special feature about Agenda 21 is therefore, that it not only points out the problem analysis (Section I) (see Figure 16.2), as well as the technical issues to solve the problem (Section II), but by means of the actors (III) and methods (IV) involved it additionally describes *the process* that leads to a sustainable model.

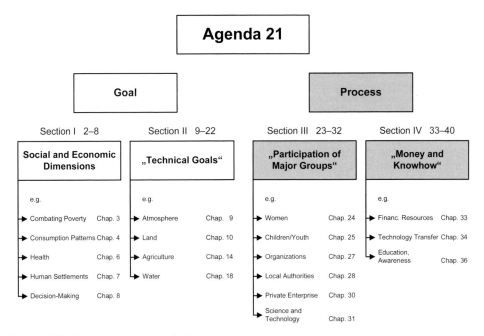

Figure 16.2 Overview over Agenda 21

A "side effect" of this participatory approach is the automatic integration of cultural aspects into the solution matrix. This can have advantages but in certain cases also disadvantages, if the almost subliminal input is not consciously processed.

As to the water medium, the implementation of this concept is best described by Integrated Water Resources Management (IWRM). In order to improve water management worldwide UNEP and the World Bank specifically created the Global Water Partnership (GWP). The IWRM explicitly performs the step from a sectoral view to an integrated view. IWRM is a highly sophisticated, well-engineered approach.

IWRM is further refined by the detailed practical instructions contained in the "tool box"[6] developed under the auspices of colleagues from The Netherlands.

An interesting assessment of the cultural factor of comprehensive structures for water management is provided by a qualitative comparison between functioning European management systems and the tool box requirements. Subject to detailed analysis, it is possible to state for Germany that while many of the conditions established within the scope of IWRM are not, or only partially complied with on the theory side, the practical results are surprisingly positive.

> *1) As a German example for imperfect structures the insufficient administrative representation of water management in river basin areas or the distribution of the responsibility (not necessarily the competences) for water among several authorities at the federal and subfederal level are to be mentioned.*
>
> *2) The economic resource assessment of Germany was criticized by the World Bank as deficient in the so-called Brisco Report, This deserves <u>careful examination</u>. According to the report, the pipeline losses are "too reduced" from an economic viewpoint because they are achieved at an excessively high cost. A remarkable opinion when considering the worldwide problem of leakage control!*

Characteristic of European water management is that 100% success is rarely achieved; in the context of worldwide benchmarking, however, the 80–90% sucess obtained is already quite desirable.

A further paradox is that worldwide it is not necessarily the countries with the strictest environmental legislation that are most successful in this field. In many regions the deficient implementation seems to be part of the political system, if not even socially consensual. Paper, particularly the one laws are written on, is obviously patient!

These examples show that the success matrix "from vision to action" involves more complex determinants. One of these determinants is culture.

16.3   Parameters of Success Under the Aspect of "Implicit Culture"

The Freshwater Conference held in Bonn discussed the conditions necessary to successfully cope with the international water problems. A hit list was established of the four most important prerequisites for a successful water policy: "1) good governance, 2) good governance, 3) good governance, 4) good governance! ". What sounds trivial is actually a challenge at all political levels, from the municipal parliament to the central government. What however is "good"?

Meanwhile, most countries have excellent environmental legislation, sometimes even better than in the benchmarking partner Europe. The legislative structures are – superficially considered – frequently similar. Differences are minimal at national and international political level. Closer to practice, at the project development and implementation levels the individual cultural particularities become increasingly strong. Additionally, state authorities and private companies operated in different ways (have different cultures) and different methods are required for successful projects. And this is not a "characteristic" of developing countries. Everywhere the features of cultural differences tend to be stronger at the base level, every group or institution develops such "distinctive" characteristics.

In the field of private enterprise this kind of phenomenon is also called corporate culture and in favourable cases is generally considered an important asset in evaluations. A widely current approach is Value-Based-Responsibility, which addresses the ethical compatibility between shareholder value and stakeholder value. The significance of corporate culture has been noted by Bringk [7] who states "Future challenge will not be technical or economic in nature, but ethical and cultural."

Cultural influences are always present within projects and they may have both positive and negative effects. This means that at least within the scope of change management [7] the culture issue needs to be deliberately addressed. Neither is it sufficient to know the "customs" of the other culture and to tolerate them "benevolently". Cultural and spiritual elements are strong and in the ideal case highly constructive!

According to international standards modern infrastructure projects contain capacity-building and awareness components. These components build the interface to the goal-oriented identification and implementation of (corporate) culture. The result should be the better fulfilment of the actual task – namely water infrastructure – through appropriate organization.

Both the technical and non-technical (organizational) issues require a highly differentiated approach. The frequently typical extremes – either random empathy and an "unconditional" acceptance of the local realities or the arrogance of taking one's own (coincidental) culture group as a model – are to be avoided. What is helpful here is that to a large extent the basic principles of sustainable development have been – at least in theory – globally perceived and accepted, as well as in many cases already transformed into laws.

The question is how (and not if) the right values and goals can be defined and successfully implemented in specific cultures, countries, businesses etc. It can be discussed how far goal-oriented implementation of this knowledge is equally anchored in all cultures (see chapter 4). Obstacles could range from religious motivations to profane "cultures" of corruption. The cultural environment has to be taken very seriously. As a matter of principle culturally and spiritually influenced positions have a natural right to exist. In one point, however, there seems to be no alternative: Cultures or societies that are unable to take and implement necessary action will fail (*To attenuate northern arrogance the reform discussion in Germany or the position of the US-Government concerning Kyoto are worth mentioning*).

16.4   Success Components According to the TTW-Model

To describe national or corporate-specific cultural aspects with discrete mathematical methods is a complex matter. For daily work purposes and after qualitative assessment of several successful national and international projects, therefore, the TTW-Project has defined four "touchstones" or levels of success. Common to all of them is that they have a "cultural factor"!

16.5   Appropriate Technology

It is generally believed that nothing is left to say about Appropriate Technology that has not already been said. Surprisingly enough, however, a look into practice tells us quite another story. The situation of wastewater treatment plants built with third-party funds worldwide, e.g. in Eastern Europe or Latin America proves this: Even without exact data it can be estimated that 80–90% of the plants do not operate satisfactorily if they operate at all. The problems even in newly built plants range from leakage control, pump dimensioning, electronic failures, up to corrosion, etc.

Sometimes it has to be feared that more systems break than are newly built. It is quite clear that this is hardly the way to achieve the MDGs. Each individual case has to be analyzed as to whether the causes are of a merely technical nature or whether culturally rooted problems concerning the handling of the technology are involved. Cultural problems can be both corporate culture problems, as well as acceptance problems. For example: Why would somebody not take care of "his" plant? Is he unable to do so (ability), is he unwilling to do so (lack of interest), or is he not allowed to do so (caste boundaries, "dirty work"). Why would somebody have so little identification with the plant encharged to him that he allows deterioration or even theft? Is money actually the only key?

Hans G. Huber, writer of chapter 17, developer for water utilities and very experienced "practical" expert, reports on an environmental assessment by UN-ECE in Ukraine: A wastewater treatment plant built with EU-funds was visited, which had only started operating six months before. The plant was not working and was beginning to decay. The reason for this was a rather insignificant detail: The rotor disc bearings of the sludge removers had obviously been dismounted quite some time previously. Would they also have been stolen if their dimensions had not been those of a current car wheel bearing?

On the other hand, what explains the counterexamples of almost lovingly tended facilities standing unexpectedly "somewhere" in "top condition"?

> *In the midst of Karalkalpakstan, a province in Usbekistan heavily affected by the Aral Sea catastrophe, a water supply plant was visited. The proud local manager spoke about "his plant" which he had painted like a work of art in the most varied colors. A closer look revealed that there was system to it: inlet pipes in blue, power in red, pretreatment in yellow, mechanical treatment in green, etc. Every time he managed to get some paint he had improved different parts of the facility. However, it was not only the color which made this plant different from many others in the region. In spite of poorest technical resources it operated well and using a number of creative solutions, was repaired time and again. By the way, parts particularly sensitive to damage and scheduled for early replacement had a brown fringe.*

Appropriate technology needs to take this kind of interfaces into consideration from the very beginning of its construction, not only to adjust itself to the "geogenic" environmental conditions but also to the anthropogenic, particularly the cultural context!

This leads us to the second major issue, which is

16.6 Management

„In spite of the not unrelevant contribution of water to prosperity worldwide, resource management is poor in most cases all over the world", observes Ismail Seralgeldin, 1992–1995 Vice-President of the World Bank. In a different context, however, he ties this in with the demand for an ample privatization of the water market as the superior management solution. As correct as he is with his criticism of worldwide water management, as careful one must be when drawing conclusions from it.

The problem is anything but new. Myriads of unsuccessful management structures in public (as well as private) enterprises, but also in projects, can be counted. The potential for performance maximization in this field is enormous.

According to what we know nowadays cultural reasons frequently prevent an absolutely proven and tested management system from being successfully transferred to another case. In addition it seems as if the managed medium obviously had a character of its own. Water management worldwide is associated with at first sight seemingly almost irrational emotions. Is it just lack of discernment when the liberalization of the electric power market finds wide acceptance but there is strong controversy when it comes to water?

- In their desperation about the great number of poorly working water projects, mostly related to state-owned utilities, the internationally development banks set their stakes mainly on the self-healing forces of the market. This was probably due to the hope that predominantly inefficient management existing in the old state-owned structures – for example "planned economy" in the model of the economy of the former USSR – would become a victim of evolution in a private business environment. This solution in its turn was influenced by a hundred-year-old banking business culture (after all banks operate in the market). But many problems emerged precisely in relation to private enterprise solutions, which – in retrospect – is not so very surprising: On the one hand the water market is strongly dependent on regional conditions, both physically as well as socioculturally. This means that the risk, and with it risk assessment, becomes extremely high and costly also in view of uncertainties in the political context.

- On the other hand, social components of water supply as a rule – and a good example of this is the supply of rural regions – hardly fit into purely market-driven solutions. "Private enterprises cannot be charitable institutions" says Gunda Röstel, former chief of the German "Green Party", now senior advisor of the private water company "Gelsenwasser", in the "Privatization" Forum at the 'Wasser Berlin 2001'. In addition to the private solution, therefore, a series of steering parameters and especially financial facilities (financial burden sharing) would be needed.

Thus it was and continues to be right to take the management issue as a starting point. Good management, however, is not a characteristic of private operators only. Much more than that, regardless of the kind of operation ownership it is a matter of internal corporate culture. Success, however, also depends on the external environment. The guiding principle is public well being, which as is generally known, is more than the sum of individual interests. Both these aspects need to be taken into account if events such as those in Cochabamba, Bolivia [8], where the conflict around water supply privatization ended in social upheaval, are to be prevented.

Likewise, there are positive examples for all types of enterprises. Such happy constellations come about when enterprises and environment (particularly also the social and legal framework) match.

Seattle Public Utilities in the US is a classic example of a worldwide disseminated type of municipally operated water services with a brilliant performance as to economic indicators, quality and sustainability criteria, as well as public awareness. Profit is made but immediately reinvested in infrastructure and quality of life for the community it serves. (This example was taken from presentations at the IWA Berlin 2000. A similarly well known example in Germany would be the Munich Public Utilities 'Stadtwerke und Stadtentwässerung München'.)

Many examples in Europe and other continents prove that Seattle is not a unique case. In particular certain state or municipal solutions are even able to cope with especially difficult conditions such as rural areas, water scarcity or high poverty rates:

> *The (state-owned) sewage disposal services of the city of Salvador (Bahia, Brazil) makes deliberate use of citizen participation in the socially disadvantaged suburbs in form of condominial solutions. Community members build (!) and maintain sewage disposal from their home connections to the sewer collector themselves, considering it common property. Condominial solutions had originally been devised for more expensive common facilities such as swimming pools, tennis courts, etc. which 'everyone would like to have but cannot normally afford'. [9]*

In the meantime the stage of "privatization at any price" as a universal remedy seems to have been partially overcome. In addition to many private enterprise approaches the example of the rehabilitation of a state-owned water utility (Uganda water supply services) was expressly quoted as a possible solution at the World Bank Water Week 2003. The number of "approved" solutions is increasing, but more important than that is to finally come closer to the intended goal.

TTW has developed a 'strategy for water infrastructure' in which control remains with the state or municipal authorities at the same time 80% of the cash flow goes to private third parties in form of service or purchase contracts [10]. The example for this model was taken from medium-sized, excellently working Bavarian facilities.

> *The Zornedinger Group, one of Bavarian AAA water suppliers located around 20 kilometers east of Munich, Bavaria, provides around 70,000 residents with a high-quality water supply. Organized as a special purpose association comprising 5 municipalities, the facilities staff consists only of a few employees. Even the maintenance is outsourced on the base of clearly defined contracts with local companies, while the spare parts deposit holds no more than a few special parts which are not easily available on the market. They deliver perfect water, 24 hours a day, 365 days per year.*

Regardless of the type of enterprise it has to be said, however, that the pursuit of better "appropriate" management at all levels is the actual prerequisite for success.

Management, though, is not a lifeless construct, but depends on people. This leads us to the third crucial factor for success:

16.7   Personalities

Existing systems worldwide need to be improved. Comprehensive improvement ultimately depends on visions and therefore on visionaries, i.e. personalities with open minds and willing to bring about change.

> *The recovery of the water supply and sewage system in Uganda was certainly not*
>
> *an easy task. W. Muhairwe, the architect of this management, wrote: "The main*
>
> *challenge of managing public enterprises lies in proper management of change".*
>
> *But … "public organisations can deliver adequate performance if well managed.*
>
> *This is clearly demonstrated by the NWSC example in which proper performance*
>
> *planning and implementation has led into performance turn- around."*
>
> *His personal share in this success can surely not be underestimated.*

That is why the human factor in its positive sense is introduced into the equation. This relates to concepts such as motivation, enthusiasm, fear, trust, lucky hand, etc., and contains a strong cultural component.

Fortunately there are a great number of examples of strong men and women capable of making things happen together with the people in their teams. Personalities are also so important because only they are able to genuinely overcome the rigid system constraints built by socially or culturally motivated obstacles. Excuses or pretexts for failure can be found everywhere. Why do such obviously clear things as the Protocol of Kyoto not work at a global level? It sometimes appears as if this "negative order of things" can not be broken. From Africa comes a comforting idea, an almost revolutionary concept if carried to its ultimate conclusion, which at the same time leads over to the next section:

Many little people at many little places,
doing many little things,
will change the face of the world

African Saying
Success Factor Networking

The fourth prerequisite is based on the personality model. In the complex network of our reality single fighters generally have no chance. Not only because single fighters fail in confrontation with the resistance capacity of organizations, but also because single personalities are no longer in a condition to capture and process the enormous amount of information necessary to bring about change. While it may sound trivial, this aspect is not always considered in project engineering.

This is why networking can be established as the fourth challenge. Behind this statement there is the perception that overproportionally many good projects or ideas fail in the real world due to communication problems. This may also sound trivial, but perhaps is not so, when the reasons for the communication issues are examined a little closer. In addition to the frequently mentioned technical reasons (lack of time, telephone didn't work, hierarchical reasons) in the background there are actually human – "all too human" – motives such as fear, isolation, poor communication skills, real or presumed political or social pressure, social obstacles, or sometimes just habits. Conversely, surprising results can be rapidly achieved when these obstacles are overcome. This is true both for the internal relations in the organization, but in particular also across organization boundaries. A purely technical model for such revolutionary obstacle jumping is e-mail communication, which across physical and social borders has created new communication codes worldwide.

A particularly important organizational example of networks and communication is participation.

The power and range of this element can be seen in the Environmental Pact Bavaria: Environmental matters and environmentally compliant attitudes are agreed upon on a fully voluntary basis between the state as supervisory and regulatory instance and private enterprises. The result is mutual commitment, as well as partly the establishment of codes of honor [11]. In the long run these processes may help create new social and cultural patterns, in addition to delivering answers which could hardly be achieved through the traditional legislative and executive channels.

Meetings of experts in greater or smaller circles, of a formal or informal kind, represent a virtual international network. Here we are also addressing the water family. This kind of communication – in the strategy devised by TTW also called 'alliance of reason' – will be one of the most positive consequences of globalization. A particular strong point related to these circles is the international matchmaking between cultures, which in the positive case can open up new solution perspectives.

Ultimately and in view of the scope of the problems the cross-sectoral thinking also demanded by IWRM will be physically viable almost exclusively in such human networks.

Cross-sectoral thinking and cross-sectoral projects are based less on formal cooperation rather than on mutual understanding and respect, at least when prompt implementation is envisaged.

Each of these four success criteria are individually highly dependent on the surrounding culture, be it either corporate or social. The cultural strengths need to be taken advantage of – an example heretofore are the strong communicative skills of the South American people. Certain corporate cultures, however, can also be obstacles to necessary change. A negative example taken from the (former?) German working culture are the often-quoted 3 defensive arguments: "We've always done it this way; this has never been done that way; and whose idea is this anyway." If this cultural background is not recognized success will be difficult to achieve.

16.8   Idea of the Implementation of Cultural Strength as an "Explicit Cultural Aim"

Culture is naturally much more than the fundament for operational projects. Culture, spirituality, religion are values in their own right. Culture also displays this value in the context of great water and sewage infrastructure projects. It seems, though, as if in spite of all theoretical knowledge this aspect is not always sufficiently taken into account. The general issue addressed here is the consideration of cultural needs and particularly the deliberate use of cultural strengths (compare also the conclusions in chapter 15). The following examples are intended to back up the basic idea:

16.9   Integral Planning

While large integral planning goals are defined in different ways (see the European Water Framework Directive or the GWP papers) it is frequently the small "loving" details which account for the quality of planning projects. For example in the case of access to watercourses: Even in areas "strictly" barred for the purpose of flood protection it is possible with certain tools to maintain access to watercourses [12]. The cultural aim is the conservation of the social function. This can only come into effect, however, if at the same time the ecological issues related to watercourse structure, water quality (wastewater systems) and leisure needs are considered. (Hardly anybody will feel like visiting a river if it is channeled in between concrete walls and smells badly). Positive consequences for the quality of life especially of low-income population groups and even social stabilization can be the reward.

Based on the guidelines of the Bavarian water management administration, flood control measures in Oberkotzau, Upper Franconia, were planned to resist a 100-year flood event. All components were devised under integral aspects: The design of the hydraulic structures was oriented by the needs of nature and town planning. The still existing narrow downtown riverbank areas were recovered, while at the same time a network of pathways – multifunctional maintenance pathways – was built according to urban needs. Until then the small town had been physically and mentally divided into two parts by the river and a railway line. A new pathway network in the central area – partly newly built in the floodplain – made the downtown bank areas attractive to people and brought the formerly separated town halves back together. Leisure areas close to town were established outside the built core zones in the remaining areas or in those regained for flood retention purposes. This also brought considerable ecological improvement.

On the left bank of the river lies the Catholic church, on the right one the Protestant church. The newly constructed path network has led to the birth of a new ecumenical citizens' action group. On the new pathway network an ecumenical trail featuring themes like water, bridges, communication and faith shall come into existence. Artworks, quotations from the Bible and other scriptures shall in future accompany strollers during their walks.

16.10 Integral Redevelopment Models

Meanwhile many international water projects have chosen the integral redevelopment approach, i.e. whole urban areas are subject to recovery both with regard to their basic sanitation systems as well as to other urban needs (urban upgrading). In spite of all efforts, however, the relationship between resources available and actual demand is quite unfavorable. Funds are extremely scarce! This poses special challenges to the integration of cultural elements:

> *A well-funded World Bank project in Pernambuco, Brazil, with a budget of US$ 89 million is responsible for the replanning of an urban area with well over 300,000 residents [13]. Half of the population lives under the UN poverty line (Family income under US$ 2 per day) in huge slums.*
>
> *The project aims to rebuild practically the entire infrastructure including dwellings and commercial facilities. With US$ 300 per person it is certainly impossible to produce a "western" or "São Paulo standard".*
>
> *An alternative presently under discussion and partially already being implemented deliberately uses the strong points of these community areas:*
>
> *Within the framework of participatory planning the needs not only of individuals but also of groups are identified. The list of basic needs such as water, food, and health is quickly followed by others like meeting and leisure places, buildings for religious services, schools, samba-dance schools, but also public order and security. What is particularly striking is the spiritual and religious orientation.*

Many of these demands have nothing or little to do with financial resources. Quality of life is purchasable only to a limited extent! Many of the successful actions are related to dignity, human dignity. Particularly the issue of crime as a problem field can hardly be overestimated. There is a culture and a spiral of crime which is hard to break through. In fact, the high crime rate in these areas significantly interferes with the execution of civil works and their later maintenance. The lack of prospects can be faced – if at all – only by "meaningful" measures. For a number of reasons the solution is unlikely to be found in consumption. Without a stronger focus on, and an accordingly stronger appraisal of intangible values and added values nothing will work. Imprecise and dangerous, however, is the adoption of a "*panem et circenses*" mentality. That is not what this is about! What is meant here is rather the reintroduction of genuine, almost quality-assured values, i.e. true cultural values.

Security issues can possibly be addressed with the help of existing social structures, particularly family relationships. The strength of women – of mothers, sisters, daughters – is known to all cultures, even if it is too seldom used. "Gender" is specially needed here!

An extremely ambitious integrated sanitation project in Brazil was threatened with failure due to the high crime rate even against the civil construction staff and workers [14]. In order to put an end to this unsustainable state of affairs the gang leaders responsible for the series of muggings and thefts were invited to a meeting in a gym. The wives, mothers and daughters who showed up in their place were very open to the changes planned in their community. Thanks to their considerable personal influence in the family they were able to bring about a "truce". As to the works themselves, also in view of the poor safety conditions they were carried out to a large extent by community residents after appropriate training.

In order to make this possible at all, however, a number of previous and in part highly sensitive measures and follow-up were required, which at a closer look reveal that in many cases the peculiarities and the culture of the people had been understood, respected and deliberately integrated into work. In line with this a community office has been established where community members can complain or make suggestions concerning water supply and sewage disposal. Every inquiry is processed and dealt with within a specific standard timeframe. People at the lower end of the social scale, who as a rule can neither read nor write, are suddenly taken seriously and are able to achieve noticeable improvement through participation! The results are remarkable. The positive outcomes are among others a small new economic cycle, a sense of ownership regarding the infrastructure facilities and a starting point for future common projects. Even the security situation improved significantly within a year.

16.11 Conclusion: Systematic Implementation of Cultural Values as Project Assets

Poverty should not be defined only in material terms. This is frequently overlooked, not only by planners of technical infrastructure measures. This can lead to problematic consequences when such a view of things prevents the recognition of the potential but also the weak points of societies. (In spite of all the cultural wealth of the "western world" sometimes the suspicion arises that wealth is understood there mainly as material values. The increasing spiritual poverty is ignored.)

The recognition of the cultural forces and demands cannot be left to sectoral considerations, but have to be communicated. In this field, cross-disciplinary cooperation is irreplaceable when it comes to the scope of scientific perception. The solution matrix, however, is only complete after the dimension of regional differences is included. The most convenient and most efficient way to get there is participation at all levels, because in the positive case it causes a certain automatic response "… what the mind won't do, the heart will accomplish …".

As a result, culture is integrated as a dimension into the Agenda 21 equation (Figure 16.3).

The cultural aims should be integrated into the impact monitoring of projects. Possible indicators could, for example, be well-functioning families, neighborhood aid, churchgoers, low crime rate, tolerance with regard to foreigners, etc. (And likewise similar indicators for the southern hemisphere).

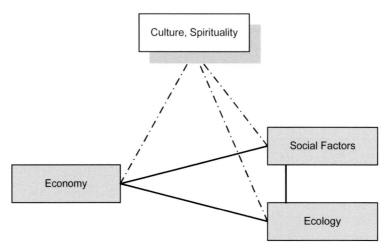

Figure 16.3 Is the famous Agenda 21 triangle in the end a tetrahedron?

Based on the experience of concrete projects the following consequences arise for an at least qualitative consideration of the implicit and explicit influences of the cultural and spiritual axiom:
- The integration of cultural aspects seems to be a basic requirement for the "from vision to action" demand
- Poverty is also spiritual (cultural) poverty
- Participation is the key to success
- At all levels – from the local project level to the global level – networks and cross-sectoral skills are required.

References

[1]   TTW, Technologietransfer Wasser, Schriften und Faltblätter des Wasserwirtschafts-amts Hof, 2002.
[2]   Wouter van Dieren, Mit der Natur rechnen, Der neue Club-of-Rome-Bericht, Birk-häuser Verlag, 1995.
[3]   F. J. Rademacher, Balance oder Zerstörung, Ökosoziales Forum Europa, 2002.
[4]   Kraemer et al, Nachhaltige Wasserwirtschaft in Deutschland, Springer Verlag, 1999.
[5]   The World Bank, Culture and Poverty, can be found under www.worldbank.org/poverty/culture, 2003.
[6]   Global Water Partnership, The Tool Box, can be found under www.gwpforum.org, 2002.
[7]   P. Senge, The Dance of Change, Doubleday, 1999.
[8]   I. Spiller, Wasser für alle?!, Heinrich Böll Stiftung, Internet, 2001.
[9]   G. da Bahia and B. Azul, Secretaria de Infra-Estrutura, Salvador (Brasilien), 2003.
[10]   M. Grambow, Strategie der vier Säulen der Wasserinfrastruktur, Presentations at ATV-DVGW Kassel, Kongressbericht, ATV-DVWK, 2001.
[11]   Bayerisches Staatsministerium für Landesentwicklung und Umweltfrage, Umweltpakt Bayern, Halbzeitbilanz, Umwelterklärung, 2003.
[12]   Bayerisches Staatsministerium für Landesentwicklung und Umweltfragen, Hochwass-erschutz in bayerischen Städten, 1998.
[13]   State of Pernambuco, Recife Upgrading Projekt, Project Appraisal Document Report No. 23331-BR, 2001.
[14]   C. Ericson Junior, Integrated Sanitation Program Mangueira & Mustardinha Suburbs, Governo do Estado Pernambuco and Prefeitura do Recife, Presentation at an AM-BITEC Seminar in Hof, 2001.

17 Sustainability from an Entrepreneurial Point of View Seen from the Particular Standpoint of a Company Active in the Water Industry

Hans G. Huber

Hans Huber AG, Industriepark Erasbach A1, 92334 Berching, Germany, hgh@huber.de

17.1 Introduction

„Act as if the maxim of thy action were to become by thy will a universal law." With this statement made in 1788, Immanuel Kant encourages each individual to show a kind of behaviour that is transferable to mankind as a whole. Immanuel Kant is asking every one of us to verify if our behaviour is consistent with the idea of sustainability.

It is commonly agreed that the idea of sustainability has become absolutely essential in every part of our lives. This is due to the increasingly intense use of the resources in our world and to an increasingly dense population on earth combined with the constant pursuit of more prosperity.

The pursuit of ever increasing prosperity is justifiable and is certainly immanent to mankind. This, in turn, makes it necessary to look for methods which are appropriate to ensure their implementation. These methods must be consistent with the criterion of sustainable action aiming to make use of our resources in such a way as to ensure their availability and to guarantee prosperity for future generations.

In this respect we must reconsider and adapt every part of our shared existence to the idea of sustainability whereas particular contributions are made by:

- Religion which provides for the ethical foundation.
- Science which provides for the theoretical foundation.
- Governance which verifies feasibility and provides for the legal foundation.

Global Sustainability. Edited by P. A. Wilderer, E. D. Schroeder, H. Kopp
Copyright © 2005 WILEY-VCH Verlag GmbH & Co. KGaA, Weinheim
ISBN: 3-527-31236-6

17.2   Role of Industry in Pursuing Sustainable Development

On the basis of ethics, science and legislation industry provides for prosperity in the material sense, i.e. implements the guidelines determined by religion, science and politics. In the light of the above, industry bears both social and ethical responsibility and cannot be excluded from the discussion on sustainability due to the broad effect of its function. Rather than discussing the issue exclusively between clergy, scientists and politicians, industry must become an integral part of the debate. Only if leaders of industry can be convinced that it is in their own interest to subscribe to sustainability can the successful implementation of its idea be ensured. We, therefore, need to draw attention of industrial circles to the necessity of sustainable action with regard to the exploitation of natural resources and the provision of well-balanced prosperity.

It is desirable to evoke this kind of awareness in industry and to promote and fund voluntary action which, in turn, must fall back upon the findings of science. The scientific approach must anticipate future developments and provide the necessary foundation required to develop theories and ideas and secure the funding needed for their implementation. Scientists can only fulfil this task by leaving the ivory tower and taking industrial concerns into account. Scientists depend not only on their own knowledge, ideas and wisdom, but on the contributions made my industry as well.

Industry acts on the basis of laws which have been prepared by politicians along the lines of sustainability. However, these laws must be understood by voters and the industrial sector if implementation is to succeed. Sustainable action in politics can only be achieved if politicians free themselves from focusing on the next election and extend their actions to development of necessary long-term approaches.

Fundamental *ethical rules* that are endorsed by religions provide a basis for sustainable development, also. Adherence to these ethical rules by industry will require open discussions by religious people, particularly religious leaders in which the grounds that have led to this ethical approach are explained.

Based upon these findings, it is essential that science, religion and politics combine their efforts with industry to enhance the idea of sustainability. They should discuss the topic together with respect for differing viewpoints and remembering that criticism can be constructive.

The question is, of course, how industry can be convinced of the need for sustainability. It must be made clear that it is in industries own best interests to adapt operations to the principles of sustainability. Thus, industry must always be interested in the prosperity of the entire population as every one of them should be regarded as a potential customer for industrial products. A well-balanced distribution of prosperity all over the world is therefore in everybody's interest. A "poor world" cannot be of interest to industry as it cannot sell anything there.

Industry must make sure that resources are not spoilt as it will need them for its own future existence. Natural resources include more than minerals such as iron ore, but also water and air. Industry must be able to act on a voluntary basis, as state-controlled regulation usually leads to unacceptable restrictions on freedom of action which, in turn, results in an attempt to avoid these imposed control measures. Any kind of self regulation by industry, however, is more likely to develop new ways and possibilities for sustainable action at every stage of the development, production and distribution of a product, as well as personnel policy. In this context positive as well as negative cases must be exposed in order to set good or bad examples.

Generally speaking, industry's commitment to sustainability can be subdivided into the following areas

1. The development of *products* and *technologies* which serve people's needs, which make life easier and which are environmentally friendly during production, use and disposal. The products must be adapted to the requirements of their application, they must be affordable and easy to handle. – They must serve people's needs.

2. Industry must commit itself to environmental protection on a voluntary basis, and subscribe to the idea of self regulation. This applies to the use of raw materials and in particular to water, air and waste. Environmental protection may be practiced by individual companies equally as well as in shared action by a group of industries in an Industrial Park. Industrial Parks play an important role in the economic development of entire continents (Asia for example). Apart from their impact on environmental protection they also point the way to the future. The underlying principle of these "Eco Industrial Estates" must be taken into account [1].

3. Sustainable company policy is an inevitable requirement for the development of a corporation as well as for the development of a region and for international business alliances. This includes sustainability in terms of marketing, market penetration and product quality. A clearly defined, sustainable business strategy is absolutely in the best interests of industry itself but is also, of course, equally beneficial to the concept of sustainability.

4. Managers should attach great significance to a sustainable personnel policy. One point is certainly to provide employment security combined with income security, but the responsibility of industry must as well include employee development and continuing education as a central issue.

5. Continuous development and adjustment in terms of sustainability should be a must for industry, which also applies of course to its products. Sustainable development is not a static thing, but a process that can only succeed if the individual living spaces and economic areas acquire and maintain a high degree of adaptability [2].

It is therefore in the interests of industry to commit itself to sustainability, and to permanently demand such sustainability from itself. This applies in particular to industries that are active in the field of water treatment.

17.3   Some Remarks about Water Industry

The sentence "Water is the basis for sustainable development" issued by UNESCO emphasises the outstanding importance of water with this sentence, and companies in this business must be especially aware of this fact. The Water Industry has therefore a special responsibility for sustainability, including its thinking and action, development of technologies, production, sales and service for supplied products. That is, the criteria that apply to industry must find particular consideration in the water industry and be made permanently visible there. Although manufacturing and contracting companies working in the Water Industry do not have the responsibility of supplying water, it is their duty to conserve water through provision of secure piped network systems and technologies for wastewater treatment including water reuse and circulation.

Water is a limited resource but is the essence of life. Sufficient and clean water must therefore be available to secure existence of man and industries. Clean water has become scarce since people increasingly agglomerate, and industries abstract clean water from nature and return polluted water (though environmental sensitivity is increasing).

Additionally, industry provides jobs, and therefore attracts people, which again increases the water consumption at industrial centres. Such concentration of people means new challenges not only for the government but also for industries, with the requirement to develop new technologies in response to the increased demand for drinking, service, and process water. It will therefore be necessary to think about using water in a different way, turning from single use to recycling.

In this context it is important to realize that regional variation of requirements must be taken into consideration: Industrialised countries usually have a sufficient amount of high-quality raw water available. Their problem and challenge is mainly treatment, recycling and returning the water to nature in an as "original" as possible condition. This also applies to by-products of wastewater treatment such as sludge. The nutrient cycle has still to be integrated and the question of phosphorus and nitrogen discharges has yet to be solved.

Developing countries, in contrast, have the problem of insufficient water availability. That is why they need different technologies for potable water treatment with particular emphasis on recycling, and also for wastewater treatment and especially recycling management. However, developing countries must be enabled to sustainably solve their problems by themselves. We can offer the problem-solving technology but these countries must be placed in a financial position to be able to pay for the solution to their problems, which will only be possible through corresponding transfers of money, job opportunities and economical prosperity. The following three guidelines should be considered:

a) Development aid payments and support for such countries are necessary.

b) It is however a better help for developing countries if industrialised countries pay an adequate price for Third World products and services. Such a transfer of money will enable developing countries to solve their water problems themselves, and also other problems (medicine, education, etc.). Developing countries are neither a resource of cheap manpower nor one of cheap products. In this context, it is indispensable to remove trade barriers which keep products from developing countries out of the markets of industrialized countries. It also implies that subsidies for the cultivation of agricultural products in developed countries are stopped so that developing countries have a chance to secure fair market prices.

c) Industrialised countries must take care not to destroy the water basis of developing countries through export of "virtual" water.

It is therefore necessary to align our research with the needs of developing countries, concentrating on products that meet their requirements for affordability and operability. To develop such technology and offer it there is a basic task of water industry. A crucial point is thereby to ensure the sustainable quality of technologies and products, combined with the necessary service plus training of operating staff in developing countries.

17.4   Water Technology Considerations

Another important point will be to provide technologies that permit water and nutrient recycling management. Although such technologies are available in theory, their implementation in practice is still inadequate. Most of all the technologies must be financially affordable, which can only be achieved through standardisation and serial production.

Irrigation technology plays an important role, too. It does not seem to make sense to waste the scarce resource of water by using irrigation systems that waste more than half the water for evaporation before it reaches the plant roots.

I am dealing with the subject of sustainability with the main focus on water deliberately from the point of view of industry as I, representing a company that specialises in products and services for the water industry, have to look at this subject of sustainability from this aspect. Another reason why I am dealing with this subject is that I am of the opinion that a shift to sustainable thinking can only be achieved if we succeed in sensitizing industry. For the awareness is absolutely there (see for instance "Econsens" – Forum for sustainable development of German economy).

Recycling of water and the simultaneous recycling of suspended material contained in it is already established technology. Figure 17.1 shows a wastewater treatment system with technologies that permit circulation of water and recovery of organic material that can be reused for application on agricultural land. This has particular importance in developing countries, where soils are generally very impoverished and un-productive.

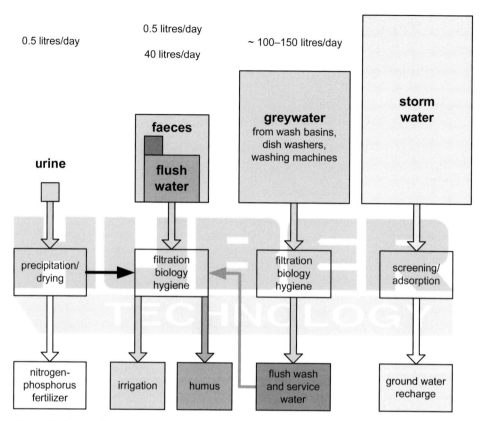

Figure 17.1 Flow diagramme of the wastewater treatment system implanted in the HUBER office building in Berching

References

[1]   A. W. König and M. Z. Wilderer in Nachhaltige Entwicklung als Herausforderung für die Entwicklungsarbeit, Metropolis Verlag 2003, pp. 199–225.

[2]   P. A. Wilderer and M. C. Wilderer in The Sustainability Axiom in Light of World Cultures 2004, Willey, VCH, Weinheim, Germany.

18   Synthesis

Peter A. Wilderer*, Edward D. Schroeder**, and Josef Bugl***

* Institute of Advanced Studies on Sustainability, European Academy of Sciences
and Arts, c/o Technical University Munich, Lehrstuhl für Wassergütewirtschaft,
Arcisstraße 21, 80333 München, Germany, peter@wilderer.de
** Department of Civil & Environmental Engineering, University of California,
1 Shields Avenue, Davis, CA 95616, USA, edschroeder@ucdavis.edu
*** Elisabeth von Thadden Str. 7, 68163 Mannheim, Germany, josefbugl@aol.com

In the following, an attempt is made to summarize the various aspects discussed in this book, and synthesize the ideas, perceptions and visions brought forward by the authors and discussed in the working groups and the plenary session.

„Sustainability" certainly contributes to the sensibility of our society towards the responsibility for the future of mankind. However, the conception of sustainability discussed so far does not have a clear target. We experience a diffusion of the definition. Thus, sustainability is used by the various groups of interest just as they need it. A prerequisite is a substantiation of the conception with the effect that it can be utilized as an instrument for the diagnosis as well as for the necessary actions to be taken.

The postulates that will result from this condensation exercise may pave the way towards a greater understanding of the interrelationships between the economic, social, ecological and cultural dimension of sustainability. They may provide a viable basis for further research and investigation, and a guideline for decision makers in science, industry and State governments.

18.1   Sustainability and Sustainable Development

Human society, its industry, governmental, and non-governmental institutions are embedded in the overall ecological system of the earth and must be understood as an integral part of the whole, not as a separate entity. As a factor in the concert of environmental factors we have the capacity to influence the ecological system of which we are a part. However, we are also influenced by changes of any other environmental factors. The human species, including its economy, can only sustain itself on earth as long as the concert of environmental factors remains favourable with respect to maintenance of reasonable living and business conditions (short term aspects), and – in the long run – control of the human population and economy.

Global Sustainability. Edited by P. A. Wilderer, E. D. Schroeder, H. Kopp
Copyright © 2005 WILEY-VCH Verlag GmbH & Co. KGaA, Weinheim
ISBN: 3-527-31236-6

Raoul Weiler states in Chapter 18 that "technological progress is considered today as an almost natural phenomenon" and "that the industrial and western society has made technological progress as its highest value, an objective to be pursued at any price". It lies in our own vital interest to define progress in a manner that will avoid unfavourable changes of the environmental system that will lead to the extinction of the human species as the ultimate result (Figure 18.1).

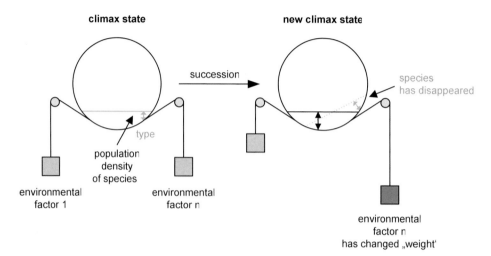

Figure 18.1 Hollow sphere model of ecological systems (adapted from Hartmann, 1960). Each point at the inner surface of the sphere represents a certain species. The sphere is partially filled with water, and the depth of the water represents the density of the respective species populations. The sphere is kept in position by a system of strings and rolls. Each string carries weights on either side representing specific environmental factors. After changing the position of one of the weights (i.e., the value of the respective environmental factor) the sphere rotates into a new position. As a result, some of the points inside of the sphere fall dry (i.e., species disappear), others become wetted (i.e., new species invade the bio-community), and the depth of the water on top of any point may change (i.e., change of the density of the respective species populations)

An ecological situation which allows man and industry to prosper may be termed "sustainable". In ecological sciences this situation is described as climax state. Climax or sustainability, respectively, is maintained as long as the weight of natural and man made ecological factors remain in a favourable range, taking into account that some of the environmental factors may not be constant but vary over time in an oscillatory manner. Sustainability is thus a dynamic state and not static in nature.

Changes of environmental factors may change the ecological situation in a habitat to the advantage of some, and to the disadvantage of other components of the ecosystem. If factors change or are changed in weight, a mostly slow succession process commences. The time during which the succession takes place depends on the specific response rate of the living elements of the ecosystem, more specifically on the replication rate of the organisms and their capability to adjust themselves to the new environmental conditions.

Long and short term changes of the environmental factors are typical for natural systems. Climatic conditions, for instance, underwent dramatic changes over the past millennia. The coming and going of the ice ages was one of the consequences and the make-up of the ecosystem changed in both cases. Similarly, we have experienced dramatic changes of the economic and political factors causing corresponding changes of the human society and economy. Preservation of an ecosystem at a particular climax state is a naïve and highly idealistic concept that may be nice to dream of but has no realistic chances of occurring, whatsoever. On the other hand, it would be disastrous to fall into a fatalistic apathy and refrain from taking active control of the ecosystem of which we are a part and that we impact through all kinds of human activities. Because we are affecting our environment, it is our responsibility to control human impacts so that future generations are able to enjoy a satisfactory habitat. Avoidance of destruction of our ecological base is to be understood as an act of self-defence by the human species.

Perfect sustainability may not be technically possible, as noted by Weber, Mosandl, and Faulstich in Chapter 1. However, approaching sustainability may be possible through careful management of our economies, as explained by Dimitris Kyriakou in Chapter 5, and by through application of developments in technology as suggested by Hans Huber (Chapter 17) may be attainable. In this context, sustainable development must be understood as an intellectual attempt to direct succession – within the boundaries over which we have control – towards a specific dynamic state that allows the human society to live in prosperity and health. Although often misunderstood, the values future generations may define as measures of prosperity and happiness are not our responsibility. We do have the responsibility, however, to avoid deterioration of our ecosystem so that future generations will survive and prosper, also.

Sustainable development is a process directed by human society, politics and industry. In affecting sustainable development we are constrained by three factors: (1) the carrying capacity of the ecosystem under consideration, (2) the variation resulting from natural changes in the environment resulting from factors such as climatic shifts, earthquakes, and volcanoes, and (3) the resiliency and adaptability of cultures. Werner Schenkel (Chapter 10) states that industrialized nations are exceeding the carrying capacity of the environment through consumption and that growth must be decoupled from consumption if goals of sustainability are to be met. Developing nations are exceeding the carrying capacity of the environment through industrialization programs and an inability to control pollution, as noted by Horst Kopp (Chapter 11) and Dietmar Rothermund (Chapter 12).

Cultural, including religious, attitudes control the way in which human societies relate to the environment. Many, if not most, of our attitudes were developed at a time when the environment was a frightening, powerful, and imponderable force and when there was minimal ecological impact by humans. Adapting to a situation in which the environment is greatly impacted by human activities has been difficult in terms of developing and following programs for sustainability. Moreover, the differences in wealth between the "haves" and "have nots" raise questions of equity in environmental matters that result in major conflicts as discussed by Armin Grunwald (Chapter 8). From this discussion two conclusions can be drawn.

1. Deterioration of our joint environment, be it on purpose or because of ignorance, must be avoided in order to establish sustainable development measures on the local, regional, and global scale.
2. Strengthening of the adaptive capacity must be a major goal of environmental politics with the aim to make our societies ready for the ever changing environmental conditions, natural or man made.

Both tasks will be difficult. Because of the adaptations that will be required, a sustainable society can only be achieved and maintained if the public is convinced of the importance of the issue. Based on history, we can assume that an educated public will accept change if there is a high level of understanding of the issues and if changes, particularly sacrifices, are shared in an equitable manner. A glance at the situation in industrialized countries reveals that the public has insufficient knowledge of the associated scientific and technical issues, which is particularly alarming since these topics are vital for the economic structures and powers of their countries. Moreover, a certain mistrust can be discerned towards decision makers in politics and industry, coupled with an increasing scepticism towards new technologies and a growing indifference towards the application of natural sciences and engineering. In order for our society to master the future problems, however, there is a strong demand, if not urgency for both, for the development and the application of sustainable technologies as well as the acceptance of responsible technologies by the public. Thus we have to enhance the public understanding and estimation of science and technology by means of improved education as outlined by Kumar in Chapter 9. Since the attitude of the public towards technological development does not only depend on scientific awareness but also to a great deal on public understanding, a specific didactic concept for knowledge transfer must be developed.

Sustainable development is a worldwide issue. Nations and regions are not so isolated as to avoid devastation by environmental degradation. However, the perspective with which people approach the issue will be strongly affected by their culture and religion, as described by Ortwin Renn in Chapter 2, Hartmann Liebetruth, in Chapter 3, and Horst Kopp in Chapter 11. Success in sustainable development will not achieved unless the perspectives of diverse cultures can be incorporated into the process. Cultural and religious diversity must become a strength of the process and sustainability must become embedded in local and regional traditions and beliefs. A principle task to be fulfilled is to develop a sensitivity for the gift of life and the natural resources in the heads and the hearts of people.

From the overall perspective, culture and cultural diversity appear thus to play an important role, and is obviously a key element of sustainability. We must see culture and cultural diversity as sources of strength and richness in incorporating alternative ways of promoting sustainability. Cultural diversity comprises local values, knowledge and time perceptions. The synergies between sustainable development strategies and expression of the local cultural heritage are of tremendous importance, and are to be strategically explored and exploited.

18.2   Postulates

Based on the consideration outlined above the participants of the workshop at Kloster Banz formulated the following postulates:

1. Sustainability cannot be achieved without eradication of poverty, and poverty eradication cannot be achieved without education. Poverty includes spiritual as well as economic poverty.
2. Measuring economic activity and quality of life with appropriate indices is necessary. Economic objectives must be balanced with sustainable ambitions.
3. Education must be based on indigenous cultural knowledge, implemented by local human resources and adapted to local present and future needs. A principle task of education is to develop sensitivity for the gift of life and the natural resources in the heads and hearts of people.
4. Culture includes religious endeavours. The principals and values common to religions worldwide – such as thankfulness for all goods on which humans depend, sensitivity for all living beings, compassion, humility and solidarity – should be utilized in formation of concepts of sustainable development.
5. Economic globalization must be based on local economic activities. Indigenous knowledge about the material and spiritual value of natural resources must be taken into consideration and adequately rewarded.
6. Sustainable development requires that local societies and economies have adaptive capacity. Local participation in planning and decision making is necessary to develop adaptive capacity. To strengthen the adaptive capacity of the various societies and economies of the world, participation methods should be further developed and rigorously implemented.
7. Science and technology is to be understood as an important means to sustainable development. Technological transfer and technological innovation must be integrated into the local cultural knowledge.

References

[1]   L. Hartmann, Die Beziehung zwischen Beschaffenheit, Leistungsfähigkeit und Lebensgemeinschaft der Belebtschlammflocke am Beispiel einer mehrstufigen Versuchsanlage (Relationship between properties, efficiency and community composition of the activated sludge floc), Vom Wasser 1960, 17, pp. 107–184.

Index

Global Sustainability. Edited by P. A. Wilderer, E. D. Schroeder, H. Kopp
Copyright © 2005 WILEY-VCH Verlag GmbH & Co. KGaA, Weinheim
ISBN: 3-527-31236-6

natural habitat
– human interference   147
neg-entropy   27, 30
negotiation   10, 119
neighbourhood aid   218
networking   16
Neumann-Morgenstern theory   77
NGOs   110, 153

O
oligopolists   73

P
participation   113, 116, 119, 141, 231
– citizen participation   211
– development of methods   231
– indigenous knowledge   166, 231
– processes   205
pedosphere   13
perspectives
– biocentric perspective   26
– proteonistic perspective   26
pesticides   149, 151
philosophy   10
planning
– cooperative planning processes   39
– environmental planning   103
– family planning   148
– integral planning   214
– participatory planning   216
policy making
– intercultural policy making   21
– neo-liberal policy   141
political subsystems   196
pollution   163, 165
– polluter pays principle   71
– pollution effects   58
population
– aging of population   59
– growth   58, 148, 198
– world population   58
poverty   45, 58, 141, 148, 154
– eradication   231
– immaterial dimension   218
pricing   70, 134

privatization   210
processes
– irreversible processes   147
– reversible processes   147
product   3
– domestic product   37
production
– net primary production   27, 30
– production factors   28, 160
– productive potential   115
– productivity   22, 32, 69

Q
quality
– quality of life   23, 27, 32, 35, 69, 190, 231
– quality of products   190
– quality of services   190
– water quality   99

R
rate
– discount rate   79
– interest rate   79
rationality
– substantive rationality   76
Rawlsion criterion   78
recycling   12, 16
region
– definition   143
– endeavours   231
regulation   28
– environmental protection regulations   45
– political regulation   113
religion   1, 6, 10
– impact of religions   131
– Islam world   56
– Judeo-Christian monotheism   55
– monotheism   55
– pagan religion   54
– religious beliefs   33, 54
– religious values   139
representation theory   162
resilience   75